MW01349240

ISO 22000 Standard Procedures for a Food Safety Management System

The Professional's Ready-to-Use Guide to Creating a Food Safety Management System for any Organization in the Supply Chain

Bizmanualz, Inc.

St. Louis, MO, USA

TRADEMARKS

Bizmanualz® is a Trademark of Bizmanualz, Inc.

This publication is sold with the understanding that the publisher is not engaged in rendering legal, accounting, or other professional services. If legal advice or other expert assistance is required, the services of a competent professional should be sought. The Publisher cannot in any way guarantee the forms, agreements and statements in this manual are being used for the purpose intended and therefore assume no responsibility for their proper and correct use.

Printed in the United States of America

ISBN-13 978-1-931591-43-0

For more information about available fully editable Policy and Procedure publications visit Bizmanualz on-line at www.bizmanualz.com.

Others in the Bizmanualz Professional's Ready-to-Use Procedure Series include:

Accounting & Bookkeeping Procedures for Internal Control
Business Security Procedures to Protect Assets and Employees
Computer & Network Procedures to Manage IT Systems
Disaster Recovery Procedures for Business Continuity Management
Finance & Treasury Procedures for Compliance and Performance
Sales & Marketing Procedures for Sales Pipeline Management
Human Resources Procedures for Employee Management
ISO 9001 Standard Procedures for a Quality Management System

Cover illustration by Leo Kundas

ISO 22000 Food Safety Management System Policies, Procedures, and Forms
Table of Contents

100 Introduction ***15***

Subject Matter Expertise 19

Section 1 – Background 23

Section 2 – ISO 9000 And ISO 22000 27

Section 3 – ISO 22000:2005 37

Section 4 – Food Safety Systems 41

Section 5 – Certification, Registration, And Accreditation 45

Section 6 – ISO Definitions 51

200 Manual Preparation ***61***

Section 1 – Introduction 67

Section 2 – Your Food Safety Manual 69

Section 3 – Effective Communication 73

Section 4 – Food Safety Procedures 79

300 Food Safety Manual ***87***

1.0 Purpose 95

2.0 Scope 95

3.0 Relation To ISO 22000:2005 95

4.0 Our Company's Food Safety Management System 95

5.0 Management Responsibility 98

6.0 Resource Management 102

7.0 Planning And Realization Of Safe Products 103

8.0 Validation, Verification, And Improvement Of The Food Safety Management System 116

FSMS Glossary 121

PROCEDURES

400 Food Safety Procedures ***129***

FS1000 – Document Control **131**

Activities
1.0 Procedure and Work Instruction Format 132
2.0 Temporary Changes 132
3.0 Document Revision. 133
4.0 Document Distribution 134
5.0 Document Control Process Review 135

Forms
FS1000-1 – Request For Document Change 137
FS1000-2 – Document Change Control 139
FS1000-3 – Document Control Database 141

FS1010 – Food Safety Records **143**

Activities
1.0 Identification of Food Safety Records 143
2.0 Food Safety Record Generation 144
3.0 Food Safety Record Maintenance 144

Forms
FS1010-1 – Food Safety Records List 147

FS1020 – Management Responsibility **151**

Activities
1.0 Management Planning 152
2.0 Management Responsibilities and Authorities 152
3.0 Management Review 153

FS1030 – Competence, Awareness, and Training **155**

Activities
1.0 New Employee Selection 155
2.0 New Employee Orientation 156
3.0 Ongoing Training 157

Forms
FS1030-1 – Food Safety Training Requirements List 159
FS1030-2 – Food Safety Training Log 161

FS1040 – Job Descriptions 163

Activities
1.0 Job Description Preparation 164
2.0 Job Description Format and Content 164
3.0 Job Description Approval and Distribution 166
4.0 Job Description Review 166

Forms
FS1040-1 – Job Description 169

FS1050 – Prerequisite Programs 171

Activities
1.0 Prerequisite Programs – Background 172
2.0 Prerequisite Program Planning 173
3.0 Implementing Prerequisite Programs 173
4.0 Reviewing Prerequisite Programs 174
5.0 Establishing Operational Prerequisite Programs 174

Forms
FS1050-1 – Prerequisite Program Example 177
FS1050-2 – Standard Operating Procedure (SOP) Form Example 179
FS1050-3 – Example Approved Chemicals / Authorized Handlers List 181
FS1050-4 – Storage Map Example 183
FS1050-5 – PRP Log Example 185

FS1060 – Hazard Analysis Preparation 187

Activities
1.0 The Food Safety Team 188
2.0 Product Characteristics 188
3.0 Intended Use 189
4.0 Flow Diagrams, Process Steps, and Control Measures 189
5.0 Verification 190

Forms
FS1060-1 – Flow Diagram Example 193

FS1070 – Hazard Analysis 195

Activities
1.0 Hazard Analysis – General Information 196
2.0 Hazard Identification and Determination of Acceptable Levels 197
3.0 Hazard Assessment 198
4.0 Selection and Assessment of Control Measures 198
5.0 Hazard Analysis Review 199

Forms
FS1070-1 – Hazard Analysis Checklist Example 201

FS1080 – HACCP Plan Management 205

Activities
1.0 Developing a HACCP Plan 206
2.0 Implementing a HACCP Plan 210
3.0 HACCP Plan Review 210
4.0 HACCP Plan Revision 211

Forms
FS1080-1 – HACCP Plan Worksheet 215
FS1080-2 – HACCP Plan Outline 217

FS1090 – Purchasing....... 219

Activities
1.0 Order Determination and Requisition 219
2.0 Order Placement 220
3.0 Recordkeeping and Matching 221
4.0 Purchasing Review 221

Forms
FS1090-1 – Purchase Requisition 223
FS1090-2 – Purchase Order 225
FS1090-3 – Purchase Order Log 227
FS1090-4 – Purchase Order Follow-Up 229

FS1100 – Supplier Evaluation 231

Activities
1.0 Approved Vendor List 232
2.0 Vendor Classification and Requirements 232
3.0 New Vendor Evaluation 233
4.0 Vendor Disqualification 234
5.0 Vendor Reevaluation 234
6.0 Second-Party Vendor Audit Program 235

Forms
FS1100-1 – Approved Vendor List 237
FS1100-2 – Vendor Survey Form 239
FS1100-3 – Approved Vendor Notification 245
FS1100-4 – Vendor Performance Log 247

FS1110 – Receiving and Inspection 249

Activities
1.0 Receiving 249
2.0 Inspection 250
3.0 Rejection, Discrepancies, and Disposition 250
4.0 Stocking 250
5.0 Receiving and Inspection Review 251

Forms
FS1110-1 – Receiving Log 253
FS1110-2 – Receiving and Inspection Report 255

FS1120 – Manufacturing 257

Activities
1.0 Kitting Work Orders 258
2.0 Production 258
3.0 Final Inspection 259
4.0 Packaging and Labeling 259
5.0 Final Release 260

Forms
None

FS1130 – Identification, Labeling, and Traceability 263

Activities
1.0 Labeling and Traceability – General 264
2.0 Labeling / Traceability System Development 264
3.0 Labeling / Tracing Food Items and Ingredients 264
4.0 Labeling / Traceability System Review 266

Forms
FS1130-1 – Lot Identification / Product Traceability Log Example 269

FS1140 – Control of Monitoring and Measuring 271

Activities
1.0 Monitoring and Measuring – General Requirements 272
2.0 Storage, Handling, and Maintenance 273
3.0 Calibration System 273
4.0 Out-Of-Tolerance Conditions 274
5.0 Control of Subcontractor Calibration 275
6.0 Test Software 275
7.0 Calibration Process Review 275

Forms
FS1140-1 – Calibration Record 277
FS1140-2 – Calibration Database 279

FS1150 – Control of Potentially Unsafe Food Product 281

Activities
1.0 General 282
2.0 Identification and Segregation of Nonconforming Product 282
3.0 Nonconformance Report 283
4.0 Evaluation for Release 283
5.0 Disposition of Potentially Unsafe Products 284

6.0 Returned Goods 284
7.0 Control Process Review 284

Forms
FS1150-1 – Nonconformance Report 287
FS1150-2 – Return Goods Authorization Form 289

FS1160 – Internal Audit and System Validation 291

Activities
1.0 Internal Audit Program 292
2.0 Internal Audit Planning 293
3.0 Conducting the Internal Audit 293
4.0 Internal Audit Reporting 295
5.0 Internal Audit Follow-Up 296
6.0 Validation of the Internal Audit Process 296

Forms
FS1160-1 – Audit Program 299
FS1160-2 – Audit Plan 301
FS1160-3 – Food Safety Audit Checklist Example 303
FS1160-4 – Final Audit Report 331

FS1170 – Corrective Action 333

Activities
1.0 Nonconformity Reports – General 333
2.0 Initiating Corrective Action 334
3.0 Investigating the Cause 335
4.0 Taking Corrective Action 335
5.0 Preventing Recurrence 335
6.0 Verification and Closure 335

Forms
FS1170-1 – Nonconformity Report 337
FS1170-2 – Corrective Action Request 339
FS1170-3 – Corrective Action Log 341

FS1180 – Continual Improvement 343

Activities
1.0 Data Collection 344
2.0 Data Analysis 345
3.0 Data Review 345
4.0 Design of Experiments 346

Forms
FS1180-1 – Variables Control Chart 349

FS1190 – Product Recall 351

Activities
1.0 Product Recall Initiation 352
2.0 Product Recall Communications...352
3.0 Handling Recalled Product 353
4.0 Review of Product Recall Process 353

Forms
FS1190-1 – Food Recall Checklist..355

FS1200 – Emergency Preparedness and Response...359

Activities
1.0 Risk Assessment and Evaluation...359
2.0 Emergency Response Planning 361
3.0 Responding to an Emergency...362
4.0 Emergency Drills and Tests 363

Forms
FS1200-1 – Risk Management Solutions Test Report 367
FS1200-2 – Emergency Response Plan...369
FS1200-3 – Emergency Response Activity Log 375

500 Index 377

[This page intentionally left blank.]

REFERENCES

1. ANSI/NCSL Z540, American National Standard for Calibration – http://www.ncslinternational.org
2. Canadian Food Inspection Agency (CFIA) –
    - Information on legislation the CFIA is responsible for enforcing – http://www.inspection.gc.ca/english/reg/rege.shtml
    - Information on fish & seafood HACCP plans – http://www.inspection.gc.ca/english/anima/fispoi/procman/sec5e.shtml
3. Developing Food Safety Systems Manual for Retail Meat Operations, Beef Information Centre, Canadian Meat Council, 2003.
4. Ensuring Safe Food – A HACCP-Based Plan for Ensuring Food Safety in Retail Establishments, Ohio State University Bulletin 901, The Ohio State University, Columbus, OH, USA – http://ohioline.osu.edu/b901/index.html.
5. EU-25 / Food and Agricultural Import Regulations and Standards – EU Traceability Guidelines 2005, USDA Foreign Agricultural Service GAIN Report, 1/21/2005.
6. FDA 306 Bioterrorism Checklist for Food Related Manufacturers, Operations Technologies, Greenville, SC, USA – http://www.operationstechnologies.com/downloads.htm.
7. Hazard Analysis And Critical Control Point Principles And Application Guidelines, National Advisory Committee On Microbiological Criteria For Foods, Food & Drug Administration, U.S. Department of Agriculture, 1997.
8. Hazard Analysis Critical Control Point (HACCP) Information Center, Iowa State University, Ames, IA, USA – http://www.iowahaccp.iastate.edu/.
9. Jorgenson, Bill, et al, FDA's Traceback Proposals Offer Challenges and Opportunities, Food Traceability Report, vol. 3, issue 6 (6/1/2003) – http://www.foodtraceabilityreport.com.
10. National Agricultural Law Center, University of Arkansas – Fayetteville, Fayetteville, AR, USA (http://www.nationalaglawcenter.org/readingrooms/foodsafety/).
11. Responding to a Food Recall, National Food Service Management Institute (NFSMI), in conjunction with the USDA Food & Nutrition Service. See http://www.nfsmi.org.
12. South Holland District Council Hazard Analysis Checklist & Advisory (http://www.sholland.gov.uk/website/pdf/eh/food/FAwardChecklistinfo.pdf).
13. Stirling Council Environmental Health – Hazard Analysis Guide (www.stirling.gov.uk/hazard_analysis_guide.doc).
14. Traceability in the Food Chain: A Preliminary Study, Food Standards Agency / ⁻ ›od Chain Strategy Division, March, 2002.
15. US FDA/CFSAN, "Guidance for Industry – Juice HACCP Hazards and Controls Guidance", 1st Edition, 2004 – http://www.cfsan.fda.gov/~dms/juicgu10.html

PROCEDURES

1. FS1000 – Document Control
2. FS1010 – Food Safety Records
3. FS1020 – Management Responsibility
4. FS1030 – Competence, Awareness, and Training
5. FS1040 – Job Descriptions
6. FS1050 – Prerequisite Programs
7. FS1060 – Hazard Analysis Preparation
8. FS1070 – Hazard Analysis
9. FS1080 – HACCP Plan Management
10. FS1090 – Purchasing
11. FS1100 – Supplier Evaluation
12. FS1110 – Receiving and Inspection
13. FS1120 – Manufacturing
14. FS1130 – Identification, Labeling, and Traceability
15. FS1140 – Control of Monitoring and Measuring
16. FS1150 – Control of Potentially Unsafe Food Product
17. FS1160 – Internal Audit and System Validation
18. FS1170 – Corrective Action
19. FS1180 – Continuous Improvement
20. FS1190 – Product Recall
21. FS1200 – Emergency Preparedness and Response

FORMS

1. FS1000-1 – Request For Document Change
2. FS1000-2 – Document Change Control
3. FS1000-3 – Document Control Database
4. FS1010-1 – Food Safety Records List
5. FS1030-1 – Food Safety Training Requirements List
6. FS1030-2 – Food Safety Training Log
7. FS1040-1 – Job Description
8. FS1050-1 – Prerequisite Program Example
9. FS1050-2 – Standard Operating Procedure (SOP) Form Example
10. FS1050-3 – Example Approved Chemicals / Authorized Handlers List
11. FS1050-4 – Storage Map Example
12. FS1050-5 – PRP Log Example
13. FS1060-1 – Flow Diagram Example
14. FS1070-1 – Hazard Analysis Checklist Example
15. FS1080-1 – HACCP Plan Worksheet
16. FS1080-2 – HACCP Plan Outline
17. FS1090-1 – Purchase Requisition
18. FS1090-2 – Purchase Order
19. FS1090-3 – Purchase Order Log
20. FS1090-4 – Purchase Order Follow-Up
21. FS1100-1 – Approved Vendor List
22. FS1100-2 – Vendor Survey Form
23. FS1100-3 – Approved Vendor Notification
24. FS1100-4 – Vendor Performance Log
25. FS1110-1 – Receiving Log
26. FS1110-2 – Receiving And Inspection Report
27. FS1130-1 – Lot Identification / Product Traceability Log Example
28. FS1140-1 – Calibration Record
29. FS1140-2 – Calibration Database
30. FS1150-1 – Nonconformance Report
31. FS1150-2 – Return Goods Authorization Form
32. FS1160-1 – Audit Program
33. FS1160-2 – Audit Plan
34. FS1160-3 – Food Safety Audit Checklist Example
35. FS1160-4 – Final Audit Report
36. FS1170-1 – Nonconformity Report
37. FS1170-2 – Corrective Action Request
38. FS1170-3 – Corrective Action Log
39. FS1180-1 – Variables Control Chart
40. FS1190-1 – Food Recall Checklist
41. FS1200-1 – Risk Management Solutions Test Report
42. FS1200-2 – Emergency Response Plan
43. FS1200-3 – Emergency Response Activity Log

[This page intentionally left blank.]

Food Safety Management System
Policies, Procedures & Forms

Section 100

Introduction

Section 100

Introduction

Introduction

A major change has recently taken place in the area of food safety. As of September, 2005, it is possible for companies to gain ISO certification for their food safety management systems. This section of the ISO 22000 manual provides an introduction to ISO 22000, a brief history of ISO, and an explanation of the process involved in certifying to ISO 22000, as well as definitions of ISO terms.

This section also provides a brief introduction to the basic concept of food safety – its structure, standards, security requirements, and definitions. The food supply chain is constantly changing, so that no document can claim to capture every possible issue, policy, or procedure and still be current. The concepts discussed in this manual cover the common, basic elements of a Food Safety Management System (FSMS).

Please read this whole section *before* you begin implementing your food safety system.

[This page intentionally left blank]

Subject Matter Expertise

JOE CARROLL, MS

Mr. Carroll is a food safety consultant with Business Edge, Ltd., in Eire (Ireland). His main experience is in the retail, food producers, and food & beverage industries. Joe has extensive experience implementing ISO 9001, HACCP, IS 343, and ISO 22000 Food Safety Management Systems. He delivers Food Hygiene and HACCP training to the food manufacturing and food service industries. Mr. Carroll holds a BS and MS in Food Science.

CHRISTOPHER ANDERSON, MBA

Christopher Anderson is currently the Managing Director of Bizmanualz, Inc., responsible for leading services engagements and directing the development of all manual products and training materials. He has over 17 years of business management and quality process consulting experience working with small to large software and technology corporations.

As founder, CEO, and President of Investorsoftware.com, an Internet catalog and e-commerce company, Chris oversaw the construction of the e-commerce website, LAN & WAN networks, firewalls, Internet marketing, and other IT systems that supported the company, including multiple Linux web, mail and e-commerce servers, Microsoft Windows 2000/NT domain controllers, PC and MAC systems, and digital copiers, faxes, and printers connected over a LAN.

Mr. Anderson served as Vice President at Arc Tangent, a software publisher, and North American distribution sales for Interactive Systems, a UNIX operating system developer, publisher and services firm. He has worked as a marketing analyst as Nixdorf Computer Corporation, electrical and computer engineer for McDonnell-Douglas Corporation (now Boeing), and served as a Lieutenant and Aeronautical Engineering Duty Officer for the U.S. Naval Air System Command. He is currently the Managing Director of Bizmanualz, Inc., and lead business process consultant and instructor for his company's "How to Create Well-Defined Processes" and "How to Align a System of People and Processes for Results" course.

Chris holds an MBA from Pepperdine University and a BSEE from Southern Illinois University-Carbondale.

STEPHEN FLICK, MBA, MIM

Steve is currently Product Manager for Bizmanualz, Inc. His responsibilities include developing new products and updating current inventory. He has over 20 years of information technology experience, working primarily with large corporations in the manufacturing and service sectors.

Steve has served a variety of organizations (including Monsanto Ag Products, Brown Group, the Federal Reserve Bank of St. Louis, the U.S. Army Reserve Personnel Center, and SBC (now AT&T)) as a programmer/analyst, systems analyst, data analyst, process analyst, data warehouse subject matter expert, and business consultant. Steve's experience and abilities as a data analyst were critical to the success of SBC Directory

Operations "win-back" campaigns. His experience in policy and procedure manuals includes developing a Corporate MIS manual for Brown Group and writing General Metal Products' ISO 9001:1994 manual.

Previously for Bizmanualz, Steve was primarily responsible for developing a Computer and Network policy and procedures manual. Steve also supervised development and production of web-based training materials for Amdocs.

Mr. Flick holds an MBA in marketing from Webster University, an MIM in information management from Washington University of St. Louis, and a BA from St. Louis University. In addition, Steve is a certified ASQ Six Sigma Green Belt and an IRCA-certified ISO 9001:2000 Auditor/ Lead Auditor.

Don Reed, MA, CQA

Don Reed develops policies, procedures, and forms for the Bizmanualz product line and creates and edits additional business materials as well. Don serves as the primary instructor for the Bizmanualz Well-Defined Processes course.

Don has extensive technical and informative writing experience. He taught Technical Communication and Advanced Technical Writing at Saint Louis University for over six years and continues to teach English Composition and Basic Writing Workshops at Southwestern Illinois College. He has taught Professional and Business Writing Seminars for numerous organizations in the St. Louis area.

Don also served as the Managing Editor of the *Journal of Policy History*, a peer-reviewed, scholarly journal published through the cooperation of the History Department at Saint Louis University and Penn State University Press.

He has twelve years of experience as a Research & Development Project Engineer, designing and building automated production equipment for a Fortune 500 manufacturer. He was intricately involved with the manufacturing processes including implementing Just-In-Time methods and achieving ISO 9001 certification. He played leadership roles in Statistical Process Control, Continual Process Improvement, Corrective Action Teams, and in creating compliance-driven documentation to achieve and maintain agency certifications.

Don has a Master of Arts in Communication from Saint Louis University and a Bachelor of Science in Electrical Engineering from Southern Illinois University in Carbondale. He is also an ASQ-certified Quality Auditor (CQA).

Introduction

Table of Contents*

JOE CARROLL, MS 3
CHRISTOPHER ANDERSON, MBA 3
STEPHEN FLICK, MBA, MIM....3
DON REED, MA, CQA 4

SECTION 1 BACKGROUND....7
International Organization for Standardization (ISO)....7
Why Register for ISO 22000? 8

SECTION 2 ISO 9000 and ISO 22000 11
ISO 9001:2000 Series Overview 11
Making the Transition from ISO 9001:2000 to ISO 22000:2005 11
ISO 22000 – ISO 9001 Cross Reference....14
Why Is There A Need For ISO 22000? 16
ISO 22000:2005 Food Safety Management Systems....16
What is Safe Food?....16
Hazard Analysis and Critical Control Points 17
The Food Supply Chain....18
Safety Guidelines and Enforcement 18
The ISO 22000 Standard 19

SECTION 3 ISO 22000:2005 21
The Design of ISO 22000....21
Hazard Analysis and Critical Control Points (HACCP)....22
An Auditable Standard 23
Copies of the ISO Standard 23
Certifying to the ISO 22000 Standard....24

SECTION 4 FOOD SAFETY SYSTEMS....25
Objectives....25
Implementation....25
Documentation 25
Level I - Quality Policies and Objectives. 26
Level II - Departmental Procedures and Responsibilities....26
Level III - Work Instructions. 27
Level IV - Forms and other Documents....27
Confidentiality Statements 27

* The Introduction Table of Contents references section page numbers (lower right corner).

SECTION 5 CERTIFICATION, REGISTRATION, AND ACCREDITATION 29
Accreditation Bodies 29
Registrars....29
Finding A Registrar....30
Selecting A Registrar 30
Auditors 30
Considerations Before Registration....31
The Registration Process 31
1. Application or Contract....32
2. Document Review 32
3. Preassessment....32
4. Assessment or Audit....32
5. Registration 33
6. Surveillances 34
Publicizing Your ISO Certification....34

SECTION 6 ISO DEFINITIONS 35

SECTION 1 BACKGROUND

Customer satisfaction, profitability, and market leadership are driven in large part by delivering safe food products to customers. Today more than ever, there is a worldwide trend towards increasingly stringent customer expectations regarding food safety. Consumers not only come to the store better educated and informed than their parents – thanks to the Internet, they can add to their knowledge about food safety almost in an instant. Accompanying this trend has been a growing realization by businesses that continual improvement is a requirement for achieving and sustaining excellent economic performance.

One roadblock to consistently providing safe food products has been the lack of a uniform set of safety standards that applied to all organizations throughout the supply chain, regardless of local laws and customs. Different countries, industries, and governments all have had varying prescriptions, standards, and systems that suppliers have had to adopt in order to deliver their goods around the world. The HACCP (hazard analysis and critical control point) With the ever increasing reach and complexity of the world's food supply chain, a single worldwide standard has been needed to simplify international commerce and ensure food safety, regardless of where products come from or where they're headed.

International Organization for Standardization (ISO)

This gave rise to ISO - the International Organization for Standardization. Located in Switzerland, ISO is the specialized international agency for standardization and the source of such standards as ISO 9001 (quality), ISO 17799 (information security), and now ISO 22000 (food safety). Established in 1947, ISO is comprised of the national standards bodies of 140 countries, working together to produce more than 13,000 international standards for business, government, and society.

ISO is made up of approximately 180 technical committees, each technical committee being responsible for one of many areas of specialization. According to ISO,

> *"The object of ISO is to promote the development of standardization and related world activities with a view to facilitating international exchange of goods and services and to developing cooperation in the sphere of intellectual, scientific, technological, and economic activity."*

The results of ISO technical work may be published as international standards.

Many people think that ISO stands for International Standards Organization. They are close, but the official name of the organization – Organisation Internationale de Normalisation – is French and translates to the International Organization for Standardization, neither of which accounts for an acronym like ISO. In fact, "ISO" is taken from the Greek word for *equal*, as in *iso*bars, *iso*sceles, and *iso*metric.

The International Organization for Standardization is an organization made up of member nations that develop standards for everything from electronics to management systems. Representatives from the member nations develop specifications and standards. The acceptance of a new or revised standard is by vote, each country getting one vote. The process itself is much more complex involving committees and subcommittees to develop and write the standards, but to accept a proposed standard, each nation has one vote and no nation outvotes another.

On the various committees, the U.S. is represented by professionals from industry, education, consulting, and registrar organizations.

Why Register for ISO 22000?

One reason to implement and certify to ISO 22000 is that some customers will require it, as they become aware of the standard and recognize the value in dealing with ISO 22000 certified suppliers. Another reason to implement the standard is its goal of harmonization – the company developing and implementing its food safety management system in conformance with ISO 22000 can be confident in its ability to conform to statutory and regulatory requirements wherever it does business.

Food companies have been developing their own HACCP plans for about a decade, following the seven HACCP principles and applying them to their circumstances in order to produce safe foods. However, HACCP plans have to be so specific to the type of business and the physical layout of each site that it is not possible to have one set of HACCP standards for all companies to follow in all situations. And while HACCP plan requirements have been codified in many localities, HACCP regulations are not – and probably cannot be – made uniform.

Companies should want to implement the ISO 22000 standard to improve their effectiveness, ensure that the products they produce are either safe to consume or will lead to the production of safe food, and increase customer satisfaction. Improved system effectiveness, production of safe food, and customer satisfaction typically result in greater profitability through gains in efficiency, effective food safety practices, and increased sales from happy customers.

Some of these benefits can be obtained by implementing and complying with the ISO 22000 standard without going through the registration process. There are benefits to be gained from registration, however. Often, when companies implement a new system, it becomes their current item of interest and soon fades with the concerns and pressures of business. It is very difficult, at times, to maintain interest and support for a system over the long term.

The certification or registration process includes regular visits by the registrar to ensure the system is maintained. Also, the outside perspective of the auditors can be extremely beneficial. A system developed and implemented internally may fall short in some areas, through no fault of the personnel involved. The objective perspective of an auditor may strengthen the system in weak areas and provide the organization with additional benefit.

Conclusion

Implementing your ISO 22000 Food Safety Management System represents a major effort. Some things will go rapidly and some will proceed in frustratingly slow fashion. It is essential that you get Top Management and all key personnel on board and not let the effort stall – it is almost always harder to get a project restarted than it is to get it started in the first place.

Although it won't seem like it at first, your ISO 22000 FSMS should ultimately provide significant benefits to your organization. The systematized continual improvement should provide efficiency gains in all areas. ISO certification of your food safety management system ought to instill greater confidence in the safety of your end products, which should translate to increased business. Improving customer satisfaction should also improve sales and, ultimately, the bottom line. Further, if an area of your program appears to be too bureaucratic and non-value-adding, it may be a target for continuous improvement efforts.

After your Company's food safety management system is ISO-certified and you've had your program in place for a year, you'll wonder how you managed without it.

[This page intentionally left blank]

SECTION 2
ISO 9000 and ISO 22000

In 1987, the first ISO 9000 standard was released. It was not until 1992, when trade throughout Europe began opening up and multinational companies in the U.S. had to meet ISO 9000 requirements in Europe, that the standard was finally implemented in the United States.

ISO 9000 has become synonymous with quality. ISO 9000 translates "quality management" into a continuously improving process designed to meet or exceed customer and regulatory requirements.

Likewise, ISO 22000 will soon become synonymous with food safety. This new ISO standard translates "food safety management" into a continuously improving process, designed to prevent or eliminate food safety hazards or, if they can't be completely eliminated, at least bring them within acceptable levels. The ISO 22000 standard was published in September, 2005.

ISO 9001:2000 Series Overview

The original set of quality assurance standards, commonly known as ISO 9000, was published in 1987 by ISO. The standard was initially based on British Standard 5750 and modified as appropriate to address issues of all member nations. The ISO 9001 series of standards are translated into many different languages and must be equal in all languages. This has resulted in the wording of the standard to be somewhat awkward at times.

The purpose of the ISO 9001 standard initially was to provide a company with the minimum requirements for a quality system to be effective in providing customers with products of a consistent quality that met their requirements. Certification to ISO 9001 provided a company's customers with confidence that the supplier had implemented an appropriate quality system and was capable of providing a product of more consistent, reliable quality. If problems with the quality of products should arise, the customer complaint and corrective action system would ensure correction of the problem and prevention of recurrence.

The standard was updated in 1994 (ISO 9001:1994) and again in December, 2000 (ISO 9001:2000).

Making the Transition from ISO 9001:2000 to ISO 22000:2005

Just as the ISO 9001 standard has addressed quality for over a decade, the new ISO 22000 standard addresses food safety. ISO 9001:1994 had gained such a level of acceptance in the world business community that at the time ISO 9001 was undergoing significant revision in 2000, the idea of developing a standard for food safety *based on* the ISO 9001 standard – one that applied to all kinds of companies and circumstances, wherever they occurred in the food supply chain – was proposed to the ISO organization.

It is for that reason that the structure and content of ISO 22000 closely resembles that of ISO 9001.

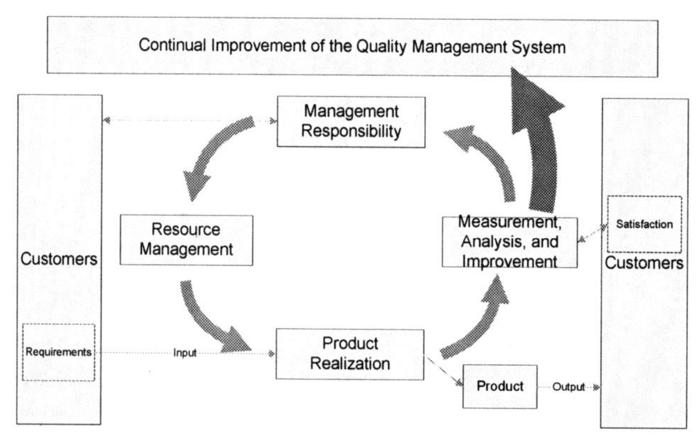

Model of a Process-Based Quality Management System (taken from ISO 9001)

The main difference in the standards – besides the overarching issue of safety versus quality (one might think safety and quality must go hand in hand, but *food* quality deals with perceptions and is, therefore, highly subjective) – is in section (clause) seven. In the ISO 9001 standard, section seven deals with planning of product realization, where the emphasis is on product design and development.

In ISO 22000, section seven is all about planning and realization of *safe* products. Instead of design and development, the emphasis is on Good Practices (GMP, GAP, etc.), prerequisite programs (PRP), and HACCP plans, all of which are designed to enable a food company to provide safe end products to its customers.

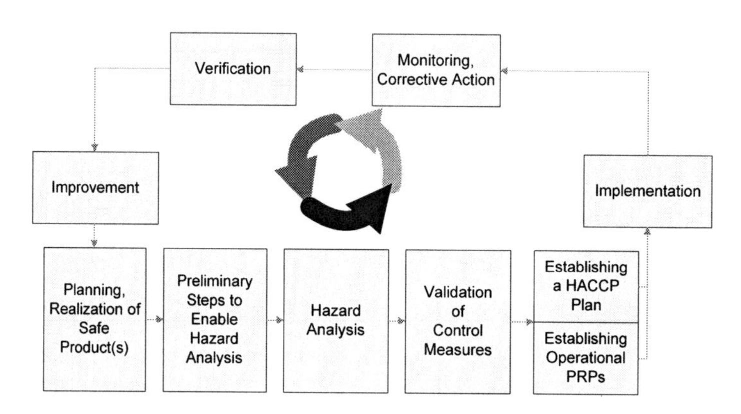

Concept of Continuous Improvement Applied to a Food Safety Management System (taken from ISO 22004)

A major benefit that comes from the two standards having a similar structure should be experienced by food companies that are already in compliance with ISO 9001. Because of the standards' common lineage, adopting the ISO 22000 food safety standard should not be difficult for companies that are ISO 9001 compliant (see the cross-reference table on the next page).

ISO 22000 – ISO 9001 Cross Reference

ISO 22000:2005			ISO 9001:2000
ISO 22000 Clause	No.	No.	ISO 9001 Clause
Introduction			**Introduction**
Scope	1	1	**Scope**
Normative References	2	2	**Normative References**
Terms and Definitions	3	3	**Terms and Definitions**
Food Safety Management System	4	4	**Quality Management System**
Management Responsibility	5	5	**Management Responsibility**
Management commitment	5.1	5.1	Management commitment
Food safety policy	5.2	5.3	Quality policy
Food safety management system planning	5.3	5.4.2	Quality management system planning
Responsibility and authority	5.4	5.5.1	Responsibility and authority
Food safety team leader	5.5	5.5.2	Management representative
Communication	5.6	5.5	Responsibility, authority, and communication
External communication	5.6.1	7.2.1 7.2.3	Determination of requirements related to product Customer communication
Internal communication	5.6.2	5.5.3 7.3.7	Internal communication Control of design & development changes
Emergency preparedness and response	5.7	5.2	Customer focus
Prerequisite programs Establishing the HACCP plan	7.2 7.6	8.5.3	Preventive action
Management review	5.8	5.6	Management review
General	5.8.1	5.6.1	General
Review input	5.8.2	5.6.2	Review input
Review output	5.8.3	5.6.3	Review output
Resource Management	6	6	**Resource Management**
Provision of resources	6.1	6.1	Provision of resources
Human resources	6.2	6.2	Human resources
General	6.2.1	6.2.1	General
Competence, awareness, and training	6.2.2	6.2.2	Competence, awareness, and training
Infrastructure	6.3	6.3	Infrastructure
Work environment	6.4	6.4	Work environment
Planning and Realization of Safe Products	7	7	**Product Realization**
General	7.1	7.1	Planning of product realization
Prerequisite programs (PRP)	7.2	6.3 6.4 7.5.1 8.5.3 7.5.5	Infrastructure Work environment Control of product and service provision Preventive action Preservation of product
Preliminary steps to enable hazard analysis	7.3	7.3	Design and development
General	7.3.1		
Food safety team	7.3.2		
Product characteristics	7.3.3	7.4.2	Purchasing requirements
Intended use	7.3.4	7.2.1	Determination of requirements related to product

ISO 22000:2005		ISO 9001:2000	
ISO 22000 Clause	**No.**	**No.**	**ISO 9001 Clause**
Flow diagrams, process steps, and control measures	7.3.5	7.2.1	Determination of requirements related to product
Hazard analysis	7.4	7.3.1	Design and development planning
General	7.4.1		
Hazard identification and determination of acceptable levels	7.4.2		
Hazard assessment	7.4.3		
Selection and assessment of control measures	7.4.4		
Establishing operational prerequisite programs	7.5	7.3.2	Design and development inputs
Establishing the HACCP plan	7.6	7.3.3	Design and development outputs
HACCP plan	7.6.1	7.5.1	Control of production and service provision
Identification of critical control points (CCP)	7.6.2		
Determination of critical limits for CCPs	7.6.3		
System for monitoring of CCPs	7.6.4	8.2.3	Monitoring and measurement of processes
Actions when monitoring results exceed critical limits	7.6.5	8.3	Control of nonconforming product
Updating preliminary information and documents specifying PRPs and the HACCP plan	7.7	4.2.3	Control of documents
Verification planning	7.8	7.3.5	Design and development verification
Traceability system	7.9	7.5.3	Identification and traceability
Control of nonconformity	7.10	8.3	Control of nonconforming product
Corrections	7.10.1	8.3	Control of nonconforming product
Corrective actions	7.10.2	8.5.2	Corrective action
Handling potentially unsafe products	7.10.3	8.3	Control of nonconforming product
Withdrawals (recalls)	7.10.4	8.3	Control of nonconforming product
Validation, Verification, and Improvement of the FSMS	**8**	**8**	**Measurement, Analysis, and Improvement**
General	8.1	8.1	General
Validation of control measure combinations	8.2	8.4 7.3.6 7.5.2	Analysis of data Design and development validation Validation of processes for production and service provision
Control of monitoring and measuring	8.3	7.6	Control of monitoring and measuring devices
FSMS verification	8.4	8.2	Monitoring and measurement
Internal audit	8.4.1	8.2.2	Internal audit
Evaluation of individual verification results	8.4.2	7.3.4 8.2.3	Design and development review Monitoring and measurement of processes
Analysis of results of verification activities	8.4.3	8.4	Analysis of data
Improvement	8.5	8.5	Improvement
Continual improvement	8.5.1	8.5.1	Continual improvement
Updating the FSMS	8.5.2	7.3.4	Design and development review

Why Is There A Need For ISO 22000?

Recent surveys conducted by various testing and governing bodies (e.g., the US Centers for Disease Control and Prevention (CDC) and the USDA Economic Research Service) suggest that our food supplies are safer than ever. However, anecdotal evidence – stories many of us have seen or heard in the national and world news in recent days and months – strongly suggests to many consumers that it isn't necessarily so.

Mad cow disease (bovine spongiform encephalopathy, or BSE) has caused countries such as Japan to ban the importing of beef. (BSE has been linked to a neurological disease in humans, known as Creutzfelt-Jakob variant.) Only after two years of rigorous, intensive testing and long negotiations between Japan and the U.S. did the Japanese market open up to American beef again.

Chronic wasting disease, which afflicts deer, elk, and moose in the wild, has become a concern for hunters and for state and federal conservation commissions. CWD, which is related to BSE, has not yet been linked to human illness but most states in the U.S. require testing of freshly harvested animals for CWD.

E. coli (in Washington state (USA), raw milk was the source of infection of at least 18 people – Dec., 2005), salmonella (about 400 people in Ontario, Canada sickened by *Salmonella*, blamed on contaminated mung bean sprouts – Dec., 2005; fresh produce may be surpassing poultry as a cause of *Salmonella*, according to CSPI – Dec., 2005), and hepatitis A (lettuce served at restaurants in Los Angeles county (Calif., USA) – Dec., 2005) have been found far more than we would like in restaurants and supermarkets.

Can consumers ever be sure that the food they eat is 100% safe? If the public needs assurances that the world food supply is safe – and they tell us they do, repeatedly – where do we begin to look for answers?

ISO 22000:2005 Food Safety Management Systems

Our concerns have been newly addressed with the September, 2005, release of the ISO 22000:2005 Food Safety Management System standard. ISO 22000 is the first international quality standard designed to work with all cultural prescriptions, statutes, and regulations. ISO 22000 is dedicated to improving consumer confidence in the food product and the process. It applies to every link in the food supply chain, from the farm to the table.

What is Safe Food?

The U.S. Food and Drug Administration (FDA) defines "safe food" as:

> *food in which illness-causing substances (bacteria, chemicals, etc.), when they are present, are within acceptable levels.*

Through food research, the definition of "acceptable levels" is continually changing. Monitoring and measuring devices are hundreds – sometimes thousands – of times more precise than those of twenty, ten, or even five years ago. What used to be measured in parts per million (PPM) can now be easily and economically measured in parts per *billion* (PPB)! Levels of food safety hazards that might have been considered safe in the 1970's or 1980's may now be considered unfit for human consumption. We are also able to find things in our food that we didn't know existed in food years ago, let alone were aware of their danger to our health.

Yet, even as we have become better informed individuals with regard to food safety, we have become more removed from the process of food production. For generations, people in developed (industrialized) countries have purchased their food from the supply chain, which has lengthened with each generation.

In 2005, the Centers for Disease Control and Prevention (CDC) reported that the U.S. experiences between 4 and 7 million cases of food-borne illness annually, resulting in 5,000 deaths and costing the economy $3-6 billion in lost production, health care, and other expenses.

As the supply chain has lengthened, the risk of encountering ***food hazards*** – any biological, chemical, or physical agent that, in the absence of any controls, is reasonably likely to cause illness or injury – has increased. Do we need more control?

Hazard Analysis and Critical Control Points

A ***critical control point*** is a step in which control can be applied and is essential to prevent or eliminate a food safety hazard or reduce it to an acceptable level, or a point or procedure in a specific food system where loss of control may result in an unacceptable health risk. I think we all would agree that hazards are waste and must be eliminated from our food supply.

Examples of food supply hazards include:

- Disease or insects;
- Contamination, pesticides, or bioterrorism.
- Mishandling or improper preparation;
- Unsanitary conditions;
- Mislabeling or improper storage;
- Transportation (not inherently a hazard but the more ingredients have to be shipped over greater distances, the greater the chance of hazard);
- Multiple governing & inspection bodies (USDA, FDA, FSIS, CDC, EPA, USDHS, NAS, NCFST, BRC, IFST, FSANZ, FSAI, ad infinitum) with plenty of opportunity for conflict because of less than optimal communication and coordination; and
- Lack of resources (money and people trained in food safety will likely always be in short supply).

Since the mid-1990's, there have been guidelines for companies in the food supply chain to follow, such as the BRC (British Retail Consortium) standard and HACCP (hazard analysis and critical control points – a set of principles, but not a true, auditable standard). Developing and following a HACCP plan makes good business sense and many governments now require that all companies in the food supply chain – from growers to sellers, and even equipment manufacturers – have their own HACCP plans.

Unfortunately, due to the varying nature of businesses in the supply chain, there has been no such thing as a HACCP plan – no uniform standard – that applied to all circumstances.

Furthermore, every nation, state, and locality has its own set of statutes and regulations governing food safety. Often, these laws are voluminous and cumbersome, and it is not unusual for food safety laws to overlap and even conflict. Up to now, there has not been a single, internationally recognized food safety standard that was applicable to every link in the supply chain and that worked in any setting, regardless of the local laws and customs. That is, not until ISO 22000.

While food safety is not guaranteed simply by virtue of a standard, with implementation and compliance with a standard like ISO 22000 throughout the food supply chain, consumers should have greater confidence in the safety and integrity of the food supply system and may be reasonably assured that the food they purchase is safe for them and their families to consume.

The Food Supply Chain

The nature of food safety, just like the nature of the food chain, has changed over the years. Generations ago, people were very close to the source of food – in the 1860's, 90% of the U.S. population was agrarian. People once raised crops and cattle for personal consumption and, when conditions permitted a bountiful harvest, sold their surplus directly to consumers at the farmers' market in the nearest town.

"Food is our common ground, a universal experience."
James Beard, 1903-1985

As the population increased and became more urbanized and the economy diversified, a food supply chain naturally came into being. Between the farmer and the consumer, companies came into being that shipped, stored, and milled grain; bagged, stored, and shipped flour; mixed flour with other ingredients to make dough; baked bread and packaged it; and stored and shipped the bread to market.

Now, people bring their cuisines and their foods with them as they move across boundaries. Production around the world increases, so that many foods are no longer seasonal items. People are generally better informed about their choices and about food safety issues. With the increasing reach and complexity of the food supply chain, food safety becomes a more important issue than ever.

Safety Guidelines and Enforcement

Food safety rules and guidelines have existed for generations but have generally been developed and applied in a piecemeal manner. Most countries have their own food safety regulations; even within countries, there is little uniformity to food safety codes. For example, the United States has thousands of federal laws pertaining to food safety and at least two departments – Agriculture and Health And Human Services – to enforce the rules.

Individual companies and food industry groups have also developed their own standards. Depending on the size and scope of a given company's business, it may be responsible for knowing and applying the rules and standards of multiple countries and industries. The plethora of food safety schemes worldwide has led to uneven application of food safety standards, confusion regarding safety requirements, and increasing complexity and costs to suppliers that are obliged to conform to multiple programs.

"If there is anything we are serious about,
it is neither religion nor learning, but food."
Lin Yutang, My Country and My People

Many governments felt a uniform set of procedures was needed; HACCP was one of the first attempts at a uniform set of guidelines. As good as HACCP is – many food companies have successfully implemented HACCP plans and many governments have passed regulations requiring food companies to develop their own HACCP plans – it isn't uniformly applicable throughout the supply chain. That is, it lends itself better to some situations than to others. This is largely how the ISO 22000 standard came into being – governments and food companies wanted a standard to which everyone could be held.

The ISO 22000 Standard

ISO 22000:2005, entitled "Food Safety Management Systems – Requirements for Any Organization in the Supply Chain", is designed to provide a framework of internationally harmonized requirements for the global food industry. It allows every type of organization in the supply chain – from primary producers to food processors, to storage and transportation companies, to retail and food service outlets, and even makers of equipment used in food processing – to implement a Food Safety Management System, or FSMS.

In addition, food safety management systems that conform to ISO 22000 can be certified, unlike with HACCP. ISO 22000 incorporates the principles of HACCP and addresses the requirements of many key standards (such as the BRC, IFS, and EU have developed) in a single source document.

A huge plus of ISO 22000 is that it parallels the ISO 9001:2000 Quality Management System standard, which has already been widely implemented in all types of industries. What this means to food companies looking to certify to ISO 22000 is that since the new standard is compatible with ISO 9001, firms that are already certified to ISO 9001 should find ISO 22000 certification relatively easy.

While food safety is not guaranteed simply by having the ISO 22000 standard, with implementation and compliance throughout the food supply chain, consumers can feel more confident that the food they buy – regardless of where it came from or how it got there – is safe to eat.

[This page intentionally left blank]

SECTION 3
ISO 22000:2005

"Food safety" is about the prevention, elimination, or control of foodborne hazards at the point of consumption. Everywhere around the world, people agree on this one point – consumers need and deserve assurance that the food sold to them is safe to consume. As food safety hazards may be introduced at any stage of the food supply chain, every company in the supply chain must exercise adequate hazard controls. In fact, food safety can only be ensured through the combined efforts of all parties in the food chain.

Organizations within the food supply chain range from primary producers (e.g., farmers, ranchers) through food processors, storage and transportation operators, subcontractors, and all the way to retail outlets (e.g., groceries, restaurants), as well as every point and company in between. And though their products are not part of the food we consume, makers of processing equipment, packaging material, cleaning agents, additives / ingredients, and even service providers (e.g., equipment testers) are also integral parts of the supply chain.

The Design of ISO 22000

ISO 22000:2005 is designed to ensure that the food supply chain has no weak links. It does this by specifying the requirements for a Food Safety Management System (FSMS) that combines the following *generally recognized key elements* to ensure food safety up to the point of the consumer:

- Interactive communication;
- System management;
- Prerequisite programs; and
- HACCP principles.

Effective communication – up and down the food supply chain – is critical to the control of food safety hazards. Communication between the Company and its suppliers – as well as between the Company and its immediate customers – helps ensure that all relevant food safety hazards are identified and are adequately controlled at each step in the supply chain. Communicating with the Company's customers and suppliers about known hazards and how to control them also helps to clarify customer and supplier requirements. All parties need to know the feasibility of and need for these requirements, as well as what their impact on the end product might be.

Understanding the Company's role and position within the food chain is an essential part of ensuring effective interactive communication throughout the chain. A simplified illustration of communication channels among members of the food supply chain is shown on the next page.

The most effective food safety system is established, operated, and updated within the framework of a structured management system and is incorporated into the Company's overall management activities, such as strategic planning. This will provide maximum benefit not just to the Company, but also to its suppliers and customers.

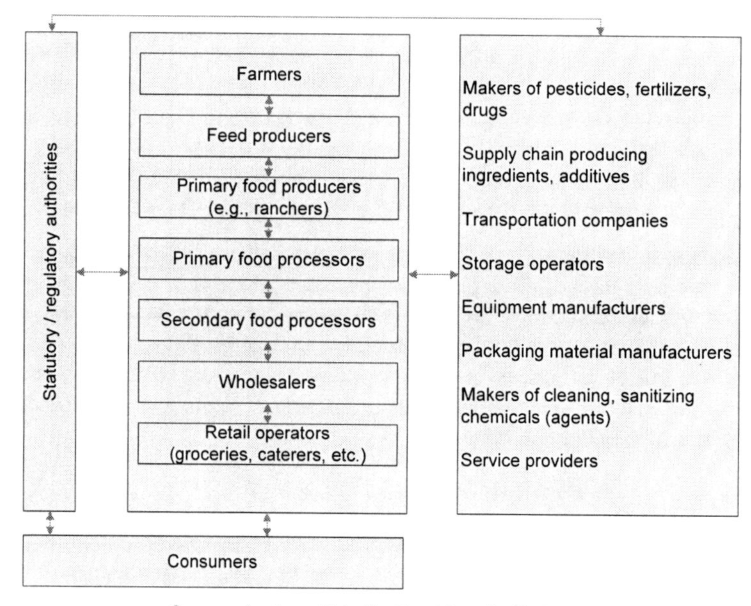

Communicating within the Food Supply Chain
(taken from ISO 22000)

ISO 22000 has been aligned with ISO 9001 to enhance the compatibility of the two standards (see the "ISO 22000 – ISO 9001 Cross Reference" table at the end of section two). Though there is naturally much commonality between the two international standards – the quality standard being a parent of the food safety standard – ISO 22000 can be applied independently of ISO 9001 or any other management system standard. The Company can align (integrate) its implementation of ISO 22000 with existing related management system requirements, or it may utilize existing management systems to establish a food safety management system that complies with ISO 22000 requirements.

Hazard Analysis and Critical Control Points (HACCP)

This standard integrates the principles of the Hazard Analysis and Critical Control Point, or HACCP, system. It also incorporates application steps developed by the Codex Alimentarius Commission, a subgroup of the World Health Organization (WHO). Furthermore, ISO 22000 combines the HACCP plan with prerequisite programs, or PRPs.

Hazard analysis is the key to every effective Food Safety Management System – a hazard analysis helps the Company acquire and organize the knowledge needed to establish and implement effective hazard control measures. ISO 22000 requires that any hazard that may be *reasonably expected to occur* in the food chain is identified and evaluated, including hazards that could be associated with the type of process and the facilities used. Thus, the food safety standard provides the means for the Company to determine and document why *it* needs to control certain identified hazards and why other companies need not.

During hazard analysis, the Company determines the strategy and methods it will use to ensure hazard control by combining PRPs, operational PRPs, and its HACCP plan(s).

An Auditable Standard

ISO 22000 was designed to be an auditable standard; by way of comparison, HACCP plans are based on the seven HACCP principles. There are guidelines and models the Company can use to develop its HACCP plans, but there is no auditable HACCP standard. It is not practical to require companies to have identical HACCP plans; in fact, any Company may need several HACCP plans, one for each product line.

One of the reasons ISO 22000 was developed as an auditable standard was to facilitate its application over the entire length of the food supply chain. The standard applies equally well to food growers, food retailers, and every step in between. Individual organizations are free to choose the necessary methods and approaches to fulfill ISO 22000 requirements, however. Companies implementing the standard should find improvements in their food safety management systems over time, regardless of whether they're certified. (To help the Company implement this standard, guidance on its use is provided in ISO/TS 22004.)

Note that ISO 22000 is intended to address food safety concerns *only*. The same approach provided by this standard may be applied to other food-related issues (e.g., ethical issues, consumer awareness).

Also noteworthy is the fact that ISO 22000 does allow smaller and less-developed companies to implement externally developed control measures.

The aim of the developers of ISO 22000 was to harmonize requirements for food safety management for businesses within the food chain on a *global* level. The standard is particularly aimed at organizations that seek a more focused, coherent, and integrated food safety management system than the law typically requires. It requires an organization to meet any applicable food safety related statutory and regulatory requirements through its food safety management system.

Copies of the ISO Standard

A copy of the standard, "ISO/FDIS 22000:2005 – Food Safety Management Systems-Requirements for any Organization in the Food Chain", is available for a small charge[1] from ISO (http://www.iso.org). You can have it shipped to you in hardcopy form or you can download it from the ISO web site in ".pdf" format.

[1] As of January 11, 2006, the cost of the standard from ISO was 120 Swiss francs, not including shipping and handling.

Certifying to the ISO 22000 Standard

BVQI (http://www.bvqi.com) was the first organization to certify a food safety management system to the ISO 22000:2005 standard; it certified a Danesco Sugar plant in late December of 2005. NSF International (http://www.nsf.org) is another organization that is capable of certifying food safety management systems. Contact ISO (http://www.iso.org) for the certifying body nearest your location.

The International Register of Certificated Auditors (IRCA – http://www.irca.org) is one organization qualified to train and certify food safety auditors. Other such organizations may be found by visiting the ISO web site.

SECTION 4
FOOD SAFETY SYSTEMS

Objectives

To assure that the Company or one of its divisions/locations will meet the requirements of ISO 22000 and achieve certification, the following processes must take place:

1. Management Decision and Commitment
2. Adequate Training and Evaluation
3. Compliance with Appropriate Standards
4. Audit and Registration

Implementation

Impediments to successful implementation of a Food Safety Management System include unrealistic time frames, resistance to change, lack of management commitment, insufficient training, and subjective interpretation of the standard.

The areas most frequently resulting in non-certification to the ISO 9001 standard, to date, have been document control, design control, purchasing, inspection and testing, quality systems, process control and inspection, and measuring and test equipment. It is conceivable that the same issues might prevent companies from certifying to ISO 22000. After all, the obstacles to ISO 9001 certification all relate to one thing – the *human factor*, which is the primary reason for having these standards in place. And although all areas of the Company's food safety program are required to be in compliance with the international standard, Company management should perform *extra* reviews to ascertain compliance in the above areas.

The appropriate personnel, under the direction of top management, should review the standards and develop, implement, and maintain a minimum set of food safety systems and procedures to satisfy the ISO 22000 standard.

Further, these personnel will provide confidence to management that the intended quality is being achieved and is:

- Documented;
- Demonstrable;
- Effective; and
- Maintained.

Documentation

Many professionals and consultants associated with ISO registration recommend organizing documentation into four levels: policies and objectives; procedures and responsibilities; work instructions; and forms and other documents.

Level I - Quality Policies and Objectives.

This first level of documentation is referred to as the "Food Safety Manual" and is separate and distinct from other procedures. The purpose of this level of documentation is to state, briefly and concisely, the Company's policies and objectives for achieving food product safety.

At a minimum, the Food Safety Manual is required to *address* each clause of the ISO 22000 standard. Each area of the Company's FSM should include three parts – scope, policy, and responsibilities:

- The **Scope** portion should simply state the purpose of the covered area;
- The **Policy** portion should state the Company's policy regarding the applicable ISO clause.
- The **Responsibility** portion should state who (by title or position, rather than by name) is responsible for the policy.

Although there is no standard format or requirement for the Food Safety Manual, a sample manual is provided in this guide for you to use as a template for creating your Company's own Manual.

Level II - Departmental Procedures and Responsibilities.

The second level of documentation should be more detailed and address the procedure(s) of an activity for a department or function and the personnel (by title or position) responsible for accomplishing the procedure(s). These procedures can be organized on a departmental basis.

Samples of Level II type departmental procedures and responsibilities documentation are provided in this guide. Food safety procedures in this manual (which ID's start with the letters "FS") are designed to be used as templates, to aid you in creating your own Company-specific procedures. The procedure format provided is not the only format or method to help you meet ISO 22000 requirements, but it has been tested and implemented over several years in different Bizmanualz products.

The procedures in this manual provide format and verbiage to describe tasks and activities that are common to organizations seeking ISO 22000:2005 requirements. They certainly are not the only format, nor are they the only functional descriptions possible, for describing your Company's tasks and activities. They are intended to provide a foundation – a framework, if you will – for you to develop procedures that satisfy your Company's – and your customers' – requirements. As you rewrite the procedures for your application, be sure to check your modifications against the requirements of the ISO standard, to ensure all requirements are addressed.

You may want to change format in the "Effectiveness Criteria" section of some procedures. For example, clause 4.1 (b) of the ISO 22000 standard requires that your Company communicate appropriate information regarding product safety issues throughout the food supply chain. It does not require or prescribe certain methods and criteria to be defined in the related procedure. *How* your Company communicates such information – and how you measure success – is up to you.

The primary reason the Effectiveness Criteria section is included in the procedures is to encourage thought in this area. Oftentimes, effectiveness of a process, task, or activity is assumed to be known (i.e., "We'll know it when we see it"). Assumptions, by their nature, are not expressed in writing and, as a result, there is little, if any, agreement between departments or even between employees in the performing department. Defining the effectiveness criteria and obtaining consensus generally results in significant improvements (in performance, attitude, and so forth).

When determining these criteria, a holistic approach as to what is best for the company should be used. An obvious example involves Purchasing. In the past, Purchasing department effectiveness was measured by the purchase price of materials. The main fallacy with this measure is that out-of-specification or marginal components result in increased costs downstream and typically result in higher total costs to the Company.

Level III - Work Instructions.

This level of documentation should be very detailed on *how* to accomplish a specific job, task, or assignment. For example, a detailed work procedure may be developed for cleaning a mixer, including step-by-step instructions on disassembly and reassembly.

Individual work instructions are very specific to industries and individual companies. Therefore, samples of level III type work instructions are not provided in this guide.

Level IV - Forms and other Documents.

The last level of documentation can include forms, records, and other documents used in the production or delivery of a product or service.

Samples of quality level IV type forms and other documents are provided in this manual and may be used as templates for you to create your own forms. The forms provided should be used as *guides*, not necessarily as final documents. For instance, a form describing a database layout is not meant to represent your Company's situation. You must determine what information your database(s) will contain and how that information is to be stored. Again, the forms are there to provide examples and foundations.

Confidentiality Statements

A *confidentiality statement* in the footer is totally optional and is recommended in order that you protect your Company's interests. Normally, when customers request a copy of your Food Safety Manual, the top-level policy manual is what they are after. The implementing procedures may contain proprietary information that you do not want to share, and usually are not meaningful to your customers.

[This page intentionally left blank]

SECTION 5 CERTIFICATION, REGISTRATION, AND ACCREDITATION

As companies become aware of the ISO 22000 standard and recognize its potential for assuring food safety throughout the supply chain, customers will expect their suppliers to have their food safety management systems *certified* (audited) or *registered* to the international standard. Certification/registration involves having an *accredited*, independent third-party registrar conduct an on-site audit of the company's operations against the requirements of the ISO standard.

The terms *certified*, *audited*, and *registered* are oftentimes used interchangeably. For ISO purposes, certification is the same as registration. Once a company has passed an *audit*, the company is *certified* by that auditor and entered into an ISO directory or *registered* for a specific ISO standard.

Therefore, accredited registrars certify organizations through an audit process for registration in an official directory of companies that have passed ISO standards.

Accreditation Bodies

Member countries have organizations that are chartered to accredit standards registrars. In the United States, this was done until recently by the Registrar Accreditation Board (RAB), now the Registration Accreditation Board/Quality Systems Association (RABQSA). RABQSA's accreditation programs for management systems certification bodies are operated by the ANSI-ASQ National Accreditation Board, or ANAB (http://www.anab.org). The auditor certification and auditor training provider programs are operated by RABQSA International (http://www.rabqsa.com). The Standards Council of Canada, or SCC (http://www.scc.ca), accredits organizations that develop standards in Canada. In the Netherlands, the Dutch Council for Accreditation (Raad voor Accreditatie, or RvA – http://www.rva.nl) is a recognized accrediting body.

ANAB, RvA, and SCC are accreditation body members of the International Accreditation Forum (IAF). A complete list of member organizations and related bodies, with links to their websites – as well as the IAF's background, goals, and policies – can be found at http://www.iaf.nu.

Registrars

Their place in the system is to certify or register companies that meet the requirements of ISO 22000:2005. Registrars must meet the requirements of the Accreditation Bodies. These requirements include things such as independence (e.g., registrars cannot provide consultation). This system ensures uniformity and impartiality in the registration process.

Finding A Registrar

Accreditation bodies maintain directories of the registrar organizations that they accredit. These directories are available on their websites. You can normally find these websites by doing a search on the accreditation body's name or initials.

Registrars in the U.S. can be found by way of the RABQSA website – http://www.rabqsa.com – which offers access to a searchable database for auditors, training courses, and accredited registrars within the United States. In Canada, the Standards Council of Canada (http://www.scc.ca) maintains a list of registrars. All other countries should consult the accreditation authority or member body for their country. Consult the ISO website for a complete list of member bodies (http://www.iso.ch/iso/en/aboutiso/isomembers/MemberList.MemberSummary?MEMBERCODE=10).

Selecting A Registrar

Registrar qualifications are a key consideration. As you research registrars, you will notice that some appear to be very limited in scope, based on their names. Registrars must be accredited in a particular industrial sector in order for them to able to certify a company in that sector. As new as the ISO 22000 standard is, your Company must be diligent in its search for a registrar. (As of early 2006, there are few qualified registrars – the number of qualified registrars can be expected to increase significantly in the very near future.) Some registrars are accredited in several, if not all, sectors; others specialize in certain sectors. The best approach to evaluating a registrar's qualifications for your Company's sector is to contact the registrar.

After qualifications, price is always a concern. Be sure to evaluate the total cost including expenses, fees and the cost of surveillances.

Probably as important as price, within limits of course, is the overall experience a client gets with a registrar. Important areas to consider are: the interpersonal skills of the auditors; the office support and its ability to get your questions answered; and whether the audit is a value-added experience (e.g., will the registrar work with you, how flexible are they in adjusting dates, how many weeks notice, etc.).

And make sure to talk to some of the registrar's clients, to evaluate their experiences and get their impressions of the registrar.

Auditors

Auditors work for or contract to registrars to perform registration assessments and surveillances – they are the front line in the process. Registrars are responsible for ensuring that auditors meet qualification requirements. Their requirements include training in auditing, specific ISO standard training, and at least one member of the audit team must have experience in the sector of the company being audited.

Verify credentials. If a person claims to be certified as an ISO 22000 lead auditor, ask to see proof of his/her certification. Make sure the certification is current by checking the expiration date. Facts about ASQ's certification programs are available at http://www.asq.org/certification/index.html.

Not all registrars require that all auditors be registered. For example, RABQSA's auditor certification program typically offers various grades of certification:

- **Provisional Auditor** is an entry grade for those with little or no management systems auditing experience. Provisional certification allows a person to participate on an audit team and gain the experience needed for advancement.
- **Auditor** certifications recognize a person's qualifications and experience to serve as a member of an audit team.
- **Lead Auditor** certification shows a person has demonstrated the ability to manage and lead an audit team.

Auditors collect objective evidence demonstrating the effectiveness (or lack thereof) of the company's management system and make registration recommendations to the registrar, the registrar having the ultimate decision in registration matters.

Considerations Before Registration

Ensure that your system is fully implemented, procedures are being followed, and records are maintained. Conduct both an internal audit and a management review, noting any exceptions before the registration process begins.

- **Internal Audits**. Use internal audits to ensure your system is in place. Make sure your records show all elements have been audited at least once.
- **Management Review**. Conduct at least one management review and ensure records (meeting minutes, for example) are maintained, showing all required items were discussed and appropriate action items assigned.
- **Exceptions**. If you have any exceptions, make sure they are noted, with appropriate rationale, in your Company's manual.

The Registration Process

The process of registration is fairly straightforward. Normally, the registrar is selected while the system is still being developed and implemented. The registrar will communicate any specific requirements that they may have or that the accreditation body may have imposed. Typically, the registrar's requirements clarify their expectations regarding implementation. For instance, a complete round of internal audits and a management review meeting must be held before the registration assessment.

Once the registrar is selected, target dates for the registration assessment are usually set. As the date approaches (usually one or two months in advance), the Registrar should be contacted to commit to firm dates. Obviously, the state of implementation of the system must be considered in setting the dates. The availability of key personnel in your company and the availability of auditors on the registrar's side both influence the final dates selected.

The number of auditors and the number of days of the registration audit depend on the size and complexity of your organization. The registrar usually communicates the total time in the quotation. This is reaffirmed and the number of auditors to be used is defined when the firm audit dates are set. When more auditors are used, the duration of the audit is shorter (for example, one auditor for four days or two auditors for two days). Keep in mind that each auditor will require an escort.

The registration process itself consists of six basic steps:

1. Application or Contract

The application also may be called a contract. It defines the rights and obligations of both parties including:

- The registrars access rights to facilities and information;
- Liability issues;
- Confidentiality;
- Rights to appeal or file a complaint;
- Instruction on use of registration certificates; and
- Conditions for terminating the application.

2. Document Review

The Registrar will require a copy of your Food Safety Manual, and implementing procedures and forms, for a document review. The document review is a desk audit of your system to verify all requirements of the standard are addressed. The Registrar will notify you of any corrections to the documentation that may be required before the Registration Assessment can be conducted.

3. Preassessment

Most registrars require a preassessment; others offer it as an option. The Pre-assessment is an initial review of the system to identify any significant omissions or weaknesses in the system and provide the organization opportunity to correct any deficiencies before the registration assessment is conducted.

NOTE: Only one preassessment may be conducted and registrars CANNOT provide consulting or advice on system implementation. Evaluating the quality system and documentation to meet ISO requirements is allowed but registrars cannot provide guidance on how to implement a quality system.

4. Assessment or Audit

In conducting an audit, ISO auditors must follow ISO 19011:2002, "Guidelines for Quality and/or Environmental Management Systems Auditing". (Despite the narrow focus implied in its name, this standard lends itself equally well to the conduct of a *food safety* audit.) The lead auditor will provide an audit schedule or plan in advance of the audit, which is meant to allow Company personnel to schedule their time appropriately and ensure they are available to the auditor(s) at the appropriate times.

In general, the flow of activities during the audit is as follows:

1. **Opening Meeting** – an introduction of the audit team and key personnel in your company. The scope and general approach to the audit is discussed. This is also

the time to question anything that is unclear in the audit schedule and communicate any last minute changes to the system or schedule.

2. **Brief tour of the facility** – keep it brief, the auditors just want to get a general feel for the layout and processes involved. This may also be done at the pre-assessment.
3. **Additional review of documents** – audit team members review documentation for areas they will audit.
4. **Examination** – the audit is conducted, personnel are interviewed, and objective evidence is collected to show the system has been effectively implemented.
5. **Daily review** – at the end of each day or the beginning of the next, the audit team reviews any issues identified during the assessment. Potential findings or nonconformities may be clarified at this time.
6. **Closing Meeting** – The audit team states their conclusions regarding the audit and presents any findings or nonconformities that were identified along with any observations they may have.
7. **Audit Report issued** – within a few weeks of the audit, the Registrar issues the audit report. The report generally restates what was discussed in the closing meeting.

During the audit, if the auditors find anything that does not meet with the requirements of the ISO standard or your procedure, they determine the severity and issue a finding. Audit findings are usually called nonconformities and fall into one of two categories depending on severity.

- A **minor nonconformity** deals with minor infractions of procedures or minor failures of the system in meeting the ISO 9001:2000 requirements. The finding of a minor nonconformity *will not* hold up your registration.
- A **major nonconformity** deals with issues where nonconforming (i.e., unsafe) product is likely to reach the customer or where there is a breakdown in the Food Safety System that results in the system not being effective in meeting the requirements of the standard. The finding of a *major* nonconformity *will* hold up your registration.

The primary difference to you between a major and minor nonconformance is that your registration cannot proceed until all major nonconformities are closed and closure is verified by the registrar. This usually involves a re-audit of the involved areas (and, of course, associated costs). Minor nonconformities require a corrective action plan and they are typically closed at the first surveillance.

5. Registration

After the audit and report are completed and all nonconformities are addressed, as appropriate, the company will receive a registration certificate that identifies the quality system as being in compliance with the ISO standard. The Company will also be listed in a register maintained by the third-party registrar. The Company is allowed to publicize this registration and use the registrar's certification mark in advertising, letterheads, and other publicity materials – the certification mark *cannot* be used on the product itself.

6. Surveillances

After your Company is registered, the process doesn't end. ISO certification is not a one-time activity. Achieving certification means that you have developed, implemented, and maintained a management system that minimally meets the requirements of the ISO standard. To ensure the system is maintained *and* that changes don't result in deficiencies in the system, registrars will perform regular surveillances of your systems.

Generally, surveillance is conducted every six months to one year (depending on the Company size and your registrar). The surveillance is similar to the registration assessment, but smaller in scope. Not as much time is spent and only a portion of the ISO standard is covered. Over the three-year period of your certificate, surveillances will eventually cover all elements of the standard. Nonconformities are handled in the same manner as in the registration assessment.

Publicizing Your ISO Certification

ISO has produced a publication, entitled *Publicizing Your ISO 9001 or ISO 14001 Certification* (and found at http://www.iso.org/iso/en/iso9000-14000/pdf/publicizing.pdf), that contains helpful information on using ISO certification for advertising or public relations. Again, though the name of this publication doesn't mention ISO 22000 and the FSMS standard is not specified in the document, the guidelines for publicizing your ISO certification are equally applicable to *any* ISO standard. Rules and restrictions apply to the use of terms and logos, so it is *important* that you read this document *before* you broadcast your certification to the public.

SECTION 6
ISO DEFINITIONS

Accreditation Act of approving an organization to operate an audit and registration program. (This is done by a nationally or internationally recognized accreditation body.)

Accreditation Body Organization chartered to accredit registrars. Accreditation bodies include RABQSA International (http://www.rabqsa.org), the Standards Council of Canada (http://www.scc.ca), and the RvA (Raad voor Accreditatie – http://www.rva.nl).

Accreditation Bodies publish the requirements that they set forth for registrars to become accredited. These requirements generally follow other ISO documents. The accreditation bodies, to ensure registrars meet and maintain systems according to the requirements, regularly audit registrars' procedures, systems, and audit practices.

Audit Standard An authentic description of essential characteristics of audits, which reflects current thought and practice.

Auditee Organization being audited.

Auditors Auditors work for or contract to registrars to perform registration assessments and surveillances. They are the "front line" in the process. Registrars are responsible for ensuring Auditors meet qualification requirements. Their requirements include training in auditing, training in the appropriate ISO standard, and at least one member of the audit team having experience in the business sector (autos, food, etc.) of the Company being audited.

Auditors collect objective evidence demonstrating the effectiveness (or lack thereof) of the company's management systems and make registration recommendations to the registrar. The registrar has the ultimate decision regarding certification.

ANSI American National Standards Institute – a private, non-profit organization that administers and coordinates the U.S. voluntary standardization and conformity assessment system.

Auditing Organization A unit or function that carries out audits through its employees. This organization may be a department of the auditee, client, or an independent third party. See "Auditors".

Capability Ability to perform designated activities and achieve results that fulfill specified requirements.

Certification Authoritative act of documenting compliance with agreed requirements.

Certification Body An impartial organization possessing the necessary competence to operate a certification program.

Characteristic Physical, chemical, visual, functional or any other identifiable property of a product, part, or material.

Client The person or organization requesting the audit. Depending on circumstances, the client may be the auditing organization, the auditee, or a third party.

Company Term used primarily to refer to a business first party, the purpose of which is to supply a product or service.

Compliance Instance of a product or service meeting the requirements of a specific standard.

Concession See "waiver".

Conformity Fulfillment of specified requirements. See "compliance".

Contract Review The systematic activities carried out before signing the contract, to ensure that requirements for quality are adequately defined, free of ambiguity, documented and realizable by the supplier.

Contractor Supplier; vendor.

Corrective Action An action taken to eliminate the causes of an existing nonconformity, defect, or other undesirable situation, to prevent recurrence. The distinction between correction such as repair, rework or adjustment and corrective action is that the former relates to the disposition of an existing nonconformity, whereas, corrective action relates to the elimination of its causes.

Criticality A relative measure of the consequences of a failure mode and its frequency of occurrence.

Customer Ultimate consumer, user, client, beneficiary, or second party.

Defect Non-fulfillment of intended usage requirements; departure or absence of one or more quality characteristics from intended usage requirements.

Degree of Demonstration The extent to which evidence is produced to provide confidence that specified requirements are fulfilled.

Deming Cycle See "Plan-Do-Check-Act".

Dependability The collective term used to describe the availability of performance and its influencing factors: reliability, performance, maintainability performance and maintenance support performance. Dependability is used only for general descriptions in non-quantitative terms and is a time-related aspect of quality.

Design Review A formal, documented, comprehensive and systematic examination of a design to evaluate the design requirements and the capability of the design to meet the requirement for quality and to identify problems and propose solutions. A design review can be conducted at any stage of the design process.

Design Specifications Description of physical and functional requirements for a product. In its initial form, the design specification is a statement of functional requirements with only general coverage of physical and test requirements. The design specification evolves through the research-and-development phase to reflect progressive refinements in performance, design, configuration, and test requirements.

Design Transfer The transfer of the design basis or baseline into specifications for the product, its components, packaging, labeling, and the manufacturing and quality assurance procedures, methods, specifications, etc., so that the product can be produced using production methods.

Discrepancy A failure to meet the specified requirement, supported by evidence (Also can be called Nonconformance, Deficiency, or Finding).

Disposition of Nonconformity The action to be taken to deal with an existing nonconforming condition in order to resolve the nonconformity.

Document Something written or printed that provides factual information or proof; any object used as evidence; to prove or support by means of documents.

Environment The conditions, circumstances, influences and stresses surrounding and affecting the product during manufacturing, storage, handling, transportation, installation and use.

Failure An event in which a previously acceptable product does not perform one or more of its required functions within the specified limits under specified conditions.

Failure Analysis The logical, systematic examination of an item, including its diagrams or formulas, to identify and analyze the probability, causes and consequences of potential and real failures.

Failure Cause The physical or chemical process, design defect, quality defect, component misapplication, or other processes which are the basic reason for failure or which initiate the physical process by which deterioration proceeds to failure.

Failure Effect The consequences a failure has on the operation, function, or status of a product.

Failure Mode The manner in which a failure is observed. The way a failure occurs and its impact on the product performance.

Failure Mode and Effects Analysis The process of identifying potential design weaknesses through reviewing schematics, engineering drawings, etc., to identify basic faults at the part/material level and determine their effect at finished or subassembly level on safety and effectiveness.

Failure Pattern The occurrence of two or more failures of the same component or feature in identical or equivalent application, which are caused by the same basic failure mechanism.

Fault Tree Analysis The process of identifying potential design weaknesses using a highly detailed logic diagram depicting basic faults and events that can lead to system failure and/or safety hazard.

ISO The International Organization for Standardization.

Lead Auditor The individual appointed by the registration organization to be responsible for the quality audit.

Management Review A formal quality evaluation, by top management, of the status and adequacy of the quality system in relation to quality policy and new objectives resulting from changing circumstances.

Nonconformity A condition of any product or component in which one or more characteristics do not conform to requirements. Includes failures, deficiencies, defects and malfunctions.

NIST The National Institute of Standards and Technology is a non-regulatory federal agency within the U.S. Commerce Department's Technology Administration. NIST's mission is to develop and promote measurement, standards, and technology to enhance productivity, facilitate trade, and improve the quality of life.

Organization A company, corporation, firm, enterprise or institution, or part thereof (whether incorporated or not, public or private) that has its own function(s) and administration that supplies products or services to other organizations. See supplier.

Organizational Structure The set of formal and informal responsibilities, authorities and relationships, arranged in a pattern, through which an organization performs its functions.

Plan-Do-Check-Act The process-based improvement cycle (also known as the "Deming Cycle") used in various ISO standards.

Policy A definite course or method of action to guide and determine present and future decisions. It is a guide to decision making under a given set of circumstances within the framework of corporate objectives, goals and management philosophies.

Preventive Action Action taken to eliminate the causes of a potential nonconformity, defect, or other undesirable situation, to prevent occurrence.

Procedure A particular way of accomplishing something, an established way of doing things, a series of steps followed in a definite regular order. It ensures the consistent and repetitive approach to actions.

Process A set of interrelated resources and activities that transform inputs into outputs with the aim of adding value. Resources include personnel, facilities, equipment, technology, methodology and finances. The aim of adding value if quality related.

Product Result of activities or processes; a product can be tangible or intangible or a combination of the two.

Product Liability A generic term used to describe the responsibility on a producer or others to make restitution for loss related to personal injury, property damage or other harm caused by a product or service. Liability is defined by law so that it may vary from country to country according to national legislation.

Production Permit A written authorization for a product, before its production, to depart from originally specified requirements; also known as deviation.

Purchaser Customer.

Qualification A documented determination that a product (and its associated software), component, packaging or labeling, meets all prescribed design and performance requirements.

Quality The composite of all the characteristics (including performance) of an item, product, or service that bear on its ability to satisfy stated or implied needs. In a contractual environment, needs are specified, whereas, in other environments, implied needs should be identified and defined. In many instances, needs can change with time; this implies periodic revision of requirements for quality. Needs are usually translated into characteristics with specified criteria. Quality is sometimes referred to as “fitness for use”, “customer satisfaction”, or “conformance to requirements.”

Quality Assurance A planned and systematic pattern of all actions necessary to provide adequate confidence that the product, its components, packaging, and labeling are acceptable for their intended use.

Quality Audit A systematic and independent examination to determine whether quality activities and related results comply with planned arrangements and whether these arrangements are implemented effectively and are suitable to achieve objectives.

The quality audit typically applies, but is not limited to a quality system or elements thereof, to processes, to products, or to services. Such audits are often called "quality system audit", "process quality audit", "product quality audit" or "service quality audit."

Quality audits are carried out by staff not having direct responsibility in the areas being audited but preferably, working in cooperation with the relevant personnel.

One purpose of a quality audit is to evaluate the need of improvement or corrective action. An audit should not be confused with surveillance or inspection activities performed for the purpose of process control or product acceptance.

Quality audits can be conducted for internal or external purposes.

Quality Audit Observation A statement of fact made during quality audit and substantiated by objective evidence.

Quality Auditor A person who has qualification status (is certified) to perform quality audits.

Quality Control The operational techniques and activities that are used to fulfill requirements for quality; all that is done to be sure that the product is what it should be.

Quality control involves operational techniques and activities aimed both at monitoring a process and at eliminating causes of unsatisfactory performance at all stages of the organization's operation in order to result in economic effectiveness.

Quality Document Document containing requirements for quality system elements for products or services; the result of activities, such as inspections or quality audits.

Quality Evaluation A systematic examination of the extent to which an entity (part, product, service or organization) is capable of meeting specified requirements. A quality evaluation may be used to determine supplier quality capability. In this case, the result of quality evaluation may be used for qualification, approval, and registration or accreditation purposes.

A quality evaluation examines potential quality capability, whereas, a quality audit additionally examines effective implementation.

Quality Improvement

The actions taken to increase the value to the customer by improving the effectiveness and efficiency of processes and activities throughout the organizational structure.

Quality Losses

The losses caused by not realizing the optimum potential of resources in processes and activities. Some examples of quality losses are the loss of customer satisfaction, loss of opportunity to add customer value, loss to the organization or society, as well as waste of resources and materials.

Quality Management

All activities of the overall management function that determine the quality policy, objectives and responsibilities, and implements them by means such as quality planning, quality control, quality assurance and quality improvement. The responsibility for quality management belongs to all levels of management but must be driven by top management. Its implementation involves all members of the organization.

Quality Management Principles

Used by management as a guide towards improving performance. The principles were derived from the experience of experts on the technical committees and represent the main elements that a good quality system must have.

The eight principles are: Customer Focus, Leadership, Involvement of People, Process Approach, Systems Approach to Management, Continual Improvement, Factual Approach to Decision-making, and Mutually Beneficial Supplier Relationship.

Quality Manual

A document stating the quality policy and describing the quality system of an organization. A quality manual may relate to the totality of an organization's activities or only to a part of it.

A quality manual will normally contain, or refer to, the quality policy, the responsibilities, authorities and inter relationships of personnel who manage, perform, verify or review work affecting quality, the quality system procedures and instruction, a statement for reviewing, updating and controlling the manual.

A quality manual can vary in depth and format to suit the needs of an organization. It may be comprised of more than one document in some instances. When its only purpose is for demonstration, it may be called a quality assurance manual.

Quality Plan A document setting out the specific quality practices, resources and sequence of activities relevant to a particular product, project or contract. A quality plan pertaining to a specific application usually references the quality manual.

Quality Policy The overall quality intentions and direction of an organization regarding quality, as formally expressed by top management.

Quality Requirements A translation of customer needs into a set of quantitatively or qualitatively stated requirements for the characteristics of a product or service to enable its realization and examination. The requirements for quality should be initially expressed in functional terms and documented.

Quality System The organizational structure, responsibilities, procedures, processes and resources for implementing quality management. The quality system should only be as comprehensive as needed to meet the quality objectives. The quality system of an organization is designed primarily to satisfy the internal requirements of the organization and is not limited to the quality assurance requirements of a particular customer. For contractual or mandatory quality assessment purposes, demonstration of the implementation of identified elements of the quality system may be required.

RABQSA International Registrar Accreditation Board / Quality Systems Accreditation, an international organization for accreditation of ISO registrars and auditor training course providers. (NOTE: RABQSA accredits food safety registrars as well as quality registrars.)

Record A document that furnishes objective evidence of activities performed or of results achieved. A quality record provides objective evidence of the extent of the fulfillment of the requirement for quality or the effectiveness of the operation of a quality system element. Some of the purposes of quality records are demonstration, traceability and corrective actions. A record can be written or stored on any data medium.

Registrar Organization that issues ISO certification; also called "certification body" or "registration body".

Their place in the system is to verify companies meet the requirements of an ISO standard (such as ISO 22000:2005) and certify or register companies that do. Registrars must meet the requirements of accreditation bodies. One key requirement is "independence" – registrars cannot give consultation in the sector in which they issue certification. This ensures uniformity and impartiality in the registration process.

Registration Formal verification by an accredited body that an organization has been audited and shown to comply with an ISO standard.

Reliability The characteristic of a product, or any component thereof, expressed as a probability that it would perform its required functions under defined conditions for specified operating periods.

Reliability Assessment A quantitative assessment of the reliability of a product, system or portion thereof. Such assessments usually employ mathematical modeling, directly applicable results of tests on the product, failure data, estimated reliability figures, and non-statistical engineering estimates.

RvA Raad voor Accreditatie, or Dutch Council for Accreditation. It was established to supervise and monitor organizations which judge quality systems and/or examine an organization on the basis of European and international standards and provide accreditation.

Sanitary and phytosanitary (SPS) regulations Government standards to protect the health of humans, plants, and animals. SPS measures are subject to World Trade Organization (WTO) rules, to prevent them from acting as non-tariff barriers.

Self-Inspection Inspection of the work performed by the performer of that work, according to specified rules. Self-inspection is used for process control by the operator.

Service The results generated by activities at the interface between the supplier and the customer and by supplier internal activities, to meet the customer requirements. Delivery or use of tangible products may form part of the service. A service may be linked with the manufacture and supply of tangible products.

Severity The consequences of a failure mode. Severity considers the worst potential consequences of a failure, determined by the degree of injury.

Specification Documented detailed requirements with which a product or service has to comply.

Supplier The organization that provides a product or service. Same as subcontractor.

System Principal entities which interact to comprise a product (e.g., hardware, software *systems*); also, an organized and disciplined approach to accomplishing a task (e.g., a failure reporting *system*).

Testing Determination, by technical or scientific means, of the properties or elements of a product or its components, including functional operation, involving the *application of* established scientific principles and procedures.

Traceability The ability to trace the history, application or location of a product and, in some cases, service by means of recorded identifications. Traceability may refer to: a product, a calibration and its relationship to the measuring equipment and the national or international standards, primary standards, basic physical constants, properties or references materials. Traceability requirements should be specified for some stated period of history or to some point of origin.

Validated Being confirmed by examination and provision of objective evidence that the particular requirements for a specific intended use have been met. Validation is normally performed on the final product under defined operating conditions, and when necessary, performed in earlier production stages. Multiple validations may be carried out if there are different intended uses.

Verification Confirmation, by examination and provision of objective evidence, that specified requirements have been met. In design and development, verification concerns the process of examining the results of a given activity to determine conformity with the input requirement for that activity.

Waiver Written authorization to release a product that does not conform to the specified requirement(s); may also be called a "concession".

Food Safety Management System Policies, Procedures & Forms

Section 200

Manual Preparation

Section 200

Manual Preparation

Manual Preparation

For the Company's food safety manual to be effective, it must be clearly written and easily understood by all employees. The objective of this manual is to certify the Company to the ISO 22000:2005 standard.

This section provides an introduction and guidance to producing your manual.

[This page intentionally left blank]

Manual Preparation
Table of Contents*

SECTION 1 INTRODUCTION 5

SECTION 2 YOUR FOOD SAFETY MANUAL 7

Style and Format: 7
Considerations In Writing Your Manual: 8
Revisions 8
Sources of Additional Information 9

SECTION 3 EFFECTIVE COMMUNICATION 11

Communication – Addressing Your Audience 11
Sexism in Writing 11
Number Usage 12
Organizing Your Thoughts 12
Outlining Technique 12
Defining the Format and Organization of Your Manual 13
Design Features 14
Style and Mechanics 14
Sources of Additional Information 15

SECTION 4 FOOD SAFETY PROCEDURES 17

Format 17
Heading Information 17
Introduction 19
The Body Of The Procedure 20
Attachments 20
Authorization 21
Production And Distribution 22
Revising And Updating Procedures 22

* The Manual Preperation Table of Contents references section page number (lower right corner).

[This page intentionally left blank]

SECTION 1
INTRODUCTION

This prototype *ISO 22000 FSMS Manual* was developed to assist organizations in preparing for ISO 22000:2005 certification. It can be custom tailored to fit one's individual company concerns and operations.

The ISO 22000 FSMS Policies, Procedures, and Forms System includes six tabbed sections consisting of the following:

100 Introduction

200 Manual Preparation

300 Food Safety Manual

400 Food Safety Procedures

500 Index

600 Notes

The *ISO 22000 FSMS* manual provides a sample food safety manual with associated procedures as a guide, but it does not intend to imply that these are the "best" or "recommended" text for your specific purposes. They can be used as a "minimum" documentation set for you to use in your effort to meet ISO 22000:2005 standards.

The language style and usage is generally representative of practices in companies based in the United States. In some cases, information may be used which is not applicable to every business.

When you edit and construct a policy, it should be easy to read, to the point, and convey a message that is clearly understood by both the employee and the management staff.

When you have completed your food safety manual and procedures, have a team of managers review them and make appropriate comments.

This publication is sold and/or distributed with the understanding that the publisher is not engaged in rendering legal, accounting, or other professional services. If legal advice or other expert assistance is required (and it is highly recommended), the services of a competent legal professional should be sought to review the final manual.

The corrected and finished product is then ready for distribution; however, it should not be considered complete. Building a Food Safety Management System is not an event – it is a continuous process of improvement, and revisions will be required from time to time.

[This page intentionally left blank]

SECTION 2
YOUR FOOD SAFETY MANUAL

This first step in building your Food Safety Management System (FSMS) is the creation of a Food Safety Manual. This is a separate and distinct step from developing your food safety procedures. The purpose is to state, in a concise and brief format, the policies and objectives of your company that are required to achieve the minimum level of food safety for your organization or division.

More than likely the input for your Food Safety Manual will come from your customers and from food safety statutes and regulations. It is concern for your customers' safety that should ultimately drive your food safety process. Their requirements, needs, and future desires are the basis for implementing your ISO 22000 food safety system in the first place.

At a minimum, your Food Safety Manual must address each one of the paragraphs of the applicable ISO Series your company plans to become registered against (ISO 22000:2005 is the focus of this manual). You may, however, wish – or need – to expand the scope of your manual to include other industry specific quality or sector requirements.

Each area covered by your FSM should include, at a minimum, three parts – scope, policy, and responsibilities:

- The "Scope" portion should simply state the purpose of the covered area;
- The "Policy" portion should state the company policy regarding the applicable ISO clause; and
- The "Responsibility" portion should state who (by generic title or position, not by name) is responsible for the policy.

ISO 22000 does not require a specific format for the Food Safety Manual. A sample manual is provided in this guide for your use as a template to create your own Food Safety Manual (see Tab 3). The Food Safety Manual's table of contents is based on the ISO 22000 standard, to ensure that each required element of the standard is addressed and provides an excellent starting point for building your FSMS.

Style and Format:

1. Use a cover or title page.
2. Include a table of contents.
3. Put policy statements on a 8 1/2" x 11 " page and print only on one side to make revisions easier.
4. Organize material by major headings, for easy reference.

5. Include an alphabetized index if your manual is lengthy.

6. Avoid a detailed paragraph identification system of numbers and letters, as this will detract from your manual's readability and the message will be lost.

7. Write in simple, easy-to-understand statements to avoid confusion.

Some companies include sample administrative forms with their procedures along with instructions for their completion. This product includes a forms section for use at your discretion in the tabbed section in the back of publication.

Considerations In Writing Your Manual:

1. It is now common practice to use pronouns that are applicable to either sex or to use his or her, or the more personal and direct, "you". Social changes influence policies on topics concerning smoking, physical fitness, etc.

2. Have your manual reviewed by an attorney, to ensure that you are in compliance with all applicable federal, state, and local laws.

3. Define terms used in your manual. Definitions should be placed in each procedure as needed also. ISO Definitions are provided in the ISO 22000 Overview section.

Revisions

Every organization is dynamic and in some state of change. This will lead to changes to your policies from time to time. Revisions should be completed and sent to all personnel who hold a copy of the manual. The revision should have an effective date and of course should be distributed in advance of the effective date. When making a change to your manual be cognizant that the language might have an indirect impact on other policies. And finally, make sure that there is a clear record of revisions made and that all employees have current information in a timely manner.

The Food Safety Manual, as well as each procedure, includes a "Revision" section at the end. It is important to keep this section, or log, up-to-date. It is the only way to ensure that distributed copies of your manual are current and approved.

Some companies, due to their size and specific business application, require expanded information. For this reason, Bizmanualz, Inc., offers additional business publications that include detailed, topic-specific manuals.

Bizmanualz, Inc., is committed to providing professional publications for those business owners and managers dedicated to the development and success of their companies. To this end, we will continue to publish useful business guides to assist you in your ever-demanding endeavors. To obtain the latest information on each of our products, visit our website at www.Bizmanualz.com.

Sources of Additional Information

With the help of the prewritten documentation contained in this publication, you should be able to produce an effective Food Safety Management System. However, you may be able to draw on other sources of information to develop a comprehensive program that truly meets the needs of your organization. These sources may include:

1. Industry or trade association publications;
2. Industry or sector consultants;
3. Your Company's legal, financial, and accounting counsel;
4. Other related company manuals and procedures;
5. Internal memos and records;
6. Equipment user manuals;
7. Customer surveys (formal or informal), proposals, or requirements;
8. Regulatory agencies (e.g., FDA, OSHA, UL);
9. Standards organizations (e.g., ISO, RABQSA, ASQ); and
10. Small Business Advisory Centers

[This page intentionally left blank.]

SECTION 3 EFFECTIVE COMMUNICATION

Communication – Addressing Your Audience

It is vital in order for your manual to be effective that it be clearly written and easily understood by all the employees in your organization. Remember that the objective of your Employee Handbook is to improve the efficiency and effectiveness of your company. Therefore, the most important rule when writing an Employee Handbook is that clarity and readability are much more important than style, perfect grammar, and a large vocabulary.

The writer must try to put him/herself in the position of the user at all times. While developing policy and procedure statements, some general guidelines for the writer to keep in mind are:

1. Explain new or unusual terms the first time they are used or in the "definitions" section of the procedure.
2. Avoid jargon wherever possible, especially when training new employees.
3. Avoid unneeded words.
4. Avoid complex writing. If the writer's vocabulary is unusually large or if they write using complex sentence structures, the writing may be at too high a level for many of the users. Understanding is far more important than the correctness of the language.
5. Use active verbs instead of passive phrases.
6. Write the way you speak. Use words and phrases that you would normally use in expressing the same thought orally.

Sexism in Writing

Webster's New Collegiate Dictionary defines "he" as used in the generic sense or when the sex of the person is unspecified. However, many people will not accept "he" being used when referring to people in general. There have been suggestions that the generic "he" be replaced with "he or she" and "him" with "him or her."

Another recent method that is gaining acceptance is the switch to a plural pronoun with a singular subject such as "When someone orders their supplies, they will have to complete a ..."

Often it is more practical for the writer to use generic nouns or by recasting sentences to include positions or titles (i.e., Applicant, Manager, Accountant, Driver, etc.) to eliminate the need for most sexism in writing of policy and procedures.

Number Usage

Writing and development of procedures may often include the frequent usage of numbers in the writing. The following rules cover how numbers should be used in print.

1. Never begin a sentence with numbers. "Fifty states have been admitted to the union" not "50 states ..."
2. Spell numbers one through nine in words, 10 or larger in numerals.
3. Compound numbers such as fractions in numerals, for example 8 1/2" X 11."
4. When showing odds (ratios), use hyphens such as 4-to-1.
5. Decimals of less than one should be preceded by a zero, for example, 0.5 percent.
6. Dates: "the 1990s"
7. Spell "%" as percent.
8. Use numerals with times and include "a.m." and "p.m."

Organizing Your Thoughts

To help the writer formulate and organize ideas for developing and writing policy and procedures, it may be useful to outline the material to be covered. Outlining is a fast and effective way to show a great amount of information in a concise, efficient manner with a minimum of writing.

To achieve well-written and easily understood policy and procedure statements that flow in a cohesive and logical form, your personnel should first outline their thoughts before beginning to write a procedure.

When outlining a policy and procedure, the following areas should be defined:

1. What is the objective that the procedure is going to accomplish?
2. What is the Company's policy on this matter?
3. Who is affected?
4. When is the policy/procedure appropriate?
5. How is the objective to be accomplished? This should include outlining major areas in a step-by-step fashion in chronological order.

Outlining Technique

There are some basic standard rules for effective and consistent outlining. It may be helpful to briefly review a few of these that deal primarily with formatting. These rules concern indentation and numbering.

Standard outline formatting is as follows:

I.
A.
1.
2.
a.
b.
B.
II.

There should always be at least two of each type of character. For example, there should not be an I without a II, an A without a B, etc.

The four characters to mark each section level, Roman numerals, capital letters, Arabic numerals and lower case letters are followed by a period. Line up the periods vertically for each type of character. Thus, Roman numerals are shown:

I.

II.

Arabic numerals are lined up as follows:

5.

39.

By aligning the periods, the text is a pleasing, straight vertical line. Indentation should be uniform, usually two to five spaces for each change in type of line. The part of a single line carried to a second line should be indented the same number of spaces as the original part of the line.

Defining the Format and Organization of Your Manual

After reviewing the preliminary listing of procedures to be included in the manual and discussing with the personnel assigned to each section, you should be able to determine the estimated length and usage of your manual. With the guidelines presented in this publication and from your own organization's needs determine the format of your policy and procedures and how you will organize the manual into sections.

The format and appearance of your policy and procedures are just as important as the organization and content of the manual. A manual that is appealing to the eye and that emphasizes the importance of the procedures is more likely to be taken seriously and used on a regular basis by employees.

However, it is important to remember that the true objective of the manual is to disseminate information in a timely and efficient manner and not to "impress" the reader with intricate headings or fancy printing techniques.

The simplest format is often the best. A simple format also allows for the most time and cost effective manner for production and maintenance of the manual. Therefore, it may be best to avoid temptations such as, detailed corporate logos in headings, two sided copies, odd sized paper patterns, expensive and restrictive binding techniques, etc.

Design Features

No Food Safety Manual or procedure should ever be regarded as "complete" in the sense that it will never change. The best manual is one that is geared to continuous growth over time and incorporates design features that make this kind of growth possible.

In this regard, use of standard 8-1/2" X 11" paper housed in a three-ring binder forms an ideal manual.

The three-ring binder provides the benefits of allowing a place for procedures to be inserted while the manual is being developed and provides for easy updating through simple replacement of pages or superseded procedures. Further, as the organization grows, the use of standard three-ring binders allows additional copies of the manual to be produced on an as needed basis instead of having to be concerned with minimum production runs required for hard bound versions.

Production of procedures on a single-side, standard size paper medium provides for easy reproduction of the manual by high-speed copiers. Use of single-side printing also provides for easy updating of the manual with changes by allowing for one or two pages to be replaced without affecting the order or sequence of the manual.

However, if the manual becomes too voluminous for ease of handling, it may be necessary to bind the manual by different sections or utilize two-sided printing to reduce paper volume.

A window type binder should be used to allow you to describe the contents of the manual on the spine, for locating sections quickly from a bookcase. The outside of the binder (front and spine) may be imprinted with the company's name and logo, to give the manual a more professional and authoritative appearance.

Divider tabs on heavy stock should be used to separate functional areas or departmental sections for ease in finding a specific procedure.

Style and Mechanics

The style and mechanics of writing include the paper, typestyle and print quality.

Paper - Some organizations use color to designate different manuals, forms, memos, etc. While color can create a pleasing appearance, it sometimes becomes too restrictive and may complicate photocopying or printing. Further, sometimes colors do not provide adequate contrast from the ink color for ease in reading. Nothing is better than black ink on white paper.

The grade of paper is usually not important since the manual is for internal purposes only and is not intended as a public relations effort. Regular 20-pound copy paper is adequate for single-sided printing. A heavier weight paper with greater opacity may be necessary if two-sided printing is used.

Typestyle - Avoid unusual artwork or type styles. These can be difficult to read and/or reproduce over a long period of time. You should strive for consistency in the overall appearance of the entire manual regardless of what area or department originates the procedures by selecting a common typewriter element or word processing typestyle font. Courier 10, Elite, 12-point Times New Roman, and 12-point Arial (or Helvetica) are common typefaces used by many business machines and word-processing software packages.

Avoid using small print, photocopy reductions, all-capital print, or fancy script styles whenever possible, as these are tiresome and difficult to read.

Adequate margins should be provided on the page. Recommended margins are a 1" top and bottom margin with a 1" left and right mirror margins, plus a 1/2" gutter margin for hole punching.

Sources of Additional Information

There are a number of books on the market that provide an excellent reference for technical writers, editors, and others developing policies and procedures. Sources include:

- Alred, Gerald, Brusaw, Charles, and Oliu, Walter, Handbook of Technical Writing, 8th Edition, St. Martin's Press, 2006 (ISBN 0-312-35267-0)
- Campbell, Nancy, Writing Effective Policies and Procedures: A Step-by-Step Resource for Clear Communication, American Management Assn., 1998 (ISBN 0-814-47960-X)
- Hartman, Peter, Starting a Documentation Group: A Hands-On Guide, Clear Point Consultants Press, 1999 (ISBN 0-967-41790-2)
- Hackos, JoAnne, Managing Your Documentation Projects, Wiley Publishing, 1994 (ISBN 0-471-59099-1)
- Microsoft Corp. Editorial Style Board, Microsoft Manual of Style for Technical Publications, 3rd Edition, 2004 (ISBN 0-735-61746-5)
- Page, Stephen, Establishing a System of Policies and Procedures, Project Management Institute, 2002 (ISBN 1-929-06500-0)
- Page, Stephen, Achieving 100% Compliance of Policies and Procedures, Process Improvement Publications, 2002 (ISBN 1-929-06549-3)

[This page intentionally left blank]

SECTION 4
FOOD SAFETY PROCEDURES

Format

Food Safety procedures address the primary steps or tasks of an activity for a department or function and the personnel (generic titles or positions) responsible for accomplishing the procedure(s). These procedures can be organized on a departmental basis.

The ISO Food Safety procedures in this manual start with the designation "FS" and may be used as a template to create your own procedures. The exact format is not prescribed by the ISO 22000 standard, though some of the field names and/or titles are required.

Heading Information

The following heading format is a compromise between simplicity and completeness. Heading information should be kept to the minimum necessary to accurately describe the procedure, identify the revision level, demonstrate authorization and be easy to produce. If you don't need all of the information called for in this format, simplify it. If you need more, add it.

A sample heading may appear as follows:

Document #: **FS1040**	Title: **JOB DESCRIPTIONS**	Print Date: **2/2/2006**
Revision #: **1.0**	Prepared By: **John Doe**	Date Prepared: **12/31/2005**
Effective Date: **2/1/2006**	Reviewed By: **Stan Middleman**	Date Reviewed: **1/12/2006**
	Approved By: **Jane Roe**	Date Approved: **1/22/2006**
Standards: **ISO 22000:2005, clause 6.2.2**		

The above heading provides the following information:

Document Number - For small manuals, no numbering system may be necessary and just the title can be used for identification purposes. With lengthy manuals, it is best to use a simple alphanumeric numbering system for identifying Procedures and for storage and retrieval purposes in data processing systems.

The first character(s) of the number are alphabetical characters that represent the subject matter covered in the manual/procedures. The letters FS have been used in these procedures to represent Food Safety. There is no obligation to retain this designation if another suits your organization better.

The remaining digits should be numeric and assigned sequentially as procedures are developed and issued.

Title – Keep document titles should be concise and descriptive. Titles are usually incorporated as part of the filename for the electronic version of the file.

Print Date – This is a field function feature of MS Word. It is useful in quickly assessing the degree to which documents used in the operation are kept up to date.

Revision Level - Once a procedure is issued, it will be subject to changes and updates as the operation matures and improves. A revision code should be used to distinguish the current document from all previous versions and assist with purging obsolete procedures. Initial procedures are often issued with a revision level of zero (0) or (0.0). If a new procedure supersedes this prior procedure, the revision number will be increased by an increment of one (1) or (.1).

Prepared By: It is useful to identify the primary individual or department that developed the procedure in the event questions arise during the approval process or subsequent to the issuance of the procedure in which additional information may be needed or to be clarified.

Date Prepared – See above

Effective Date - The actual date the procedure or revision will be implemented. Note the effective date is <u>not</u> the issue date of the original procedure. The effective date should be in the format of MM/DD/YY and is located directly beneath the procedure number and revision level.

Reviewed By – Multiple review levels may not be needed for simple procedures, but good procedures can be improved with the addition of another set of eyes. This field provides a record, if necessary, of additional review.

Date Reviewed – See above

Approved By - After the procedure has been properly reviewed and authorized, the title page is initialed by an authorized individual (typically, this is someone in top management, like the chief financial officer or the product vice president). It is best to initial the approval section by hand rather than type it in; this clearly indicates the PROCEDURE has been authorized and distinguishes the final version from any draft versions that may still be in circulation. The "Approved by" identification should be placed directly underneath the "Prepared by" section.

Page Numbering - All pages should be numbered in the form Page # of # to quickly identify the order or placement of pages within a procedure. It is also useful when updating an existing procedure where only one or two pages will be replaced. The page number forms part of the footer for each page.

In addition to the page number, the title page and all subsequent pages are identified in the footer by the complete filename used for the electronic version of the document. This filename includes the document number, title and revision level.

Introduction

Located under the title block discussed above, the introduction section provides the information necessary for the reader to determine the Company policy covering this area, the purpose of the procedure, whom it affects and in what situation(s), and the definition of any new or unusual terms.

A sample Introduction section can appear as follows:

Policy: It is the policy of Sample Company that all departments will prepare and maintain standardized operating policies and procedures that cover the performance of all major functions within their department.

Purpose: This procedure outlines the steps involved in preparing, maintaining and approving standard operating policies and procedures in order to provide consistent, informative and effective procedures to the employees of Sample Company.

Scope: This procedure applies to all policies and procedures used or written by all departments and individuals of Sample Company.

Responsibilities:

All Personnel are required to understand and use this procedure.

Management is responsible for maintaining this procedure.

Definitions: Policy - A definite course or method of action to guide and determine present and future decisions. It is a guide to decision making under a given set of circumstances within the framework of corporate objectives, goals and management philosophies.

Procedure - A particular way of accomplishing something, an established way of doing things, a series of steps followed in a definite regular order. It ensures the consistent and repetitive approach to actions.

A general description of the Introduction is as follows:

Policy - The policy should clearly indicate the company's or top management's beliefs or protocol affecting this area.

Purpose - A brief description of the objective of the procedure, it should expand and clarify the Policy statement.

Scope - Describes the areas, functions, individuals, or departments affected by the procedure and in what situations the procedure applies.

Responsibilities - Describe who, in generic titles or positions, is responsible for implementing or maintaining the procedure or parts of the procedure. For example, write "The Department Manager is responsible for...", *not* "John Doe is responsible for...".

Definitions - Describes any terms contained in the procedure that may be new or unusual to the reader.

The Body Of The Procedure

The body of the procedure includes a complete description of the policy and/or the procedure, the methods to be used, form names, cross references to other procedures related to it, etc.

Although the narrative information in the body can vary considerably in format, it is imperative that it clearly explains to the reader in an orderly fashion exactly how to accomplish the objective of the procedure.

For lengthy procedures, it may be useful to identify and segregate the steps or areas by a numbering system. For example, each procedural category would be identified with a step number, starting with 1.0 and a heading description. All the steps within the category would then be numbered sequentially (For example, 1.1, 1.2, 1.3, etc.). The step number should be located at the left margin with the related narrative indented.

In addition to the detailed steps for implementing the procedure, a well written procedure will include the following items:

Effectiveness Criteria - Describe any thresholds or standards that are used to evaluate the work product or results of the procedure. How will an employee know that the procedure was executed correctly?

References - List applicable documents, procedures, manuals, laws and regulations, validation studies, or other sources that were used to develop, refine, or influence the policy/procedure statement.

Records - Describe the records, minutes, reports, notes, forms, or other documents that are generated or used when implementation the procedure.

Revision History - Describe all revisions made to the procedure. Include a revision number, date of the revision, a short description of the changes made, and a source of the request for the change.

Attachments

Any forms, diagrams, illustrations or other documents referenced in the body of the procedure should be attached and referenced as exhibits in the Records section using a sequential numbering system, (For example, Exhibit 1, Exhibit 2, etc.). It may be useful to use copies of actual completed forms or documents as exhibits to illustrate to the reader how they are completed.

Once all comments have been received and a final version approved, the procedure should be printed in its final form. The procedure should then be authorized by the appropriate individual and released for production and distribution.

Authorization

Origination of policies and procedures usually begins at the unit level by employees or department managers. Once a draft copy of a proposed procedure is developed it should be reviewed, corrected if necessary and approved before being released as a corporate policy and procedure. The approval process generally consists of review for consistency and accuracy, conflict with corporate policy or with other procedures and general readability.

The approval process can vary widely between companies but it is recommended to keep approvals to a minimum if possible. If too many people or managers of equal ranking are required to authorize a procedure, it can turn the development of procedures into a bureaucratic process, which can considerably slow the release of policies and procedures while adding little value to the final version.

A method for gaining the input of others while streamlining the process and keeping authority at the functional or departmental level is to release draft copies of proposed procedures to a select number of individuals for comment. It should be made clear to these individuals that they should confine their suggestions to what they feel is really essential to the procedure's accuracy, readability, and usefulness.

The selection of who should receive a draft copy will depend primarily on the nature and content of the procedure. Sensitive issues or areas that deal with corporate exposure such as in the personnel area, intellectual property or trade secrets should be reviewed by top management including the president and may also include the company's legal counsel.

Rudimentary procedures that affect only a small unit within the company and are likely to be of no interest to others should be kept to a minimal review process. However, it is still advisable to have someone familiar with the area but separate or outside of the unit or department review the proposed procedure. For example, the company's finance officer may review a proposed accounting SOP. This type of review serves three purposes:

- First, what makes sense to the preparer directly involved in enforcing the policy or conformance to a procedure may not be understood when read for the first time by someone not as closely involved;
- Second, a review by multiple department managers may prevent a conflict with a policy/procedure document still in the formative or discussion stage elsewhere in the Company of which the original preparer was unaware; and
- Third, it allows the input of multiple individuals while allowing the department manager to maintain control of the integrity of the procedure and drive its completion and release in a timely manner.

Production And Distribution

Once a policy and procedure statement is authorized, it may be duplicated on standard white copy paper and three-hole-punched on the left margin. Multiple page procedures may be corner stapled to prevent losing pages until it is included in the manual.

The number of copies of the procedure should correlate to the number of manuals that have been distributed. It is generally advisable to designate one individual in the company for the production of procedures. This individual keeps track of the number of manuals issued and ensures that new or revised procedures are distributed to the appropriate personnel.

Management of the company should decide which departments or positions will receive copies of the manual or as an alternative, sections of the manual may be distributed which pertain only to a specific function or department. However, if the manual is to serve as a communication tool, enough copies of the manual should be available to employees.

Since the manual contains many operating procedures that are vital to the company's business practices and methods, there should be some accountability for the manuals. Generally, one individual will maintain a list of the number of copies in circulation and the names of those to whom they have been assigned. When a supervisor or manager leaves the company, there should be a strong incentive to return their copy of the manual; some companies withhold final compensation until the manual has been returned.

However, one should avoid numbering each copy of a manual, unless absolutely necessary. Numbering implies confidentiality and some degree of importance, which may not be the case. Besides, the issuer must maintain a permanent record of the numbers cross-referenced to the recipients that can make personnel changes a tedious record-keeping task.

The Company may elect to produce, distribute, and control copies of its manual electronically. According to the laws of most countries, electronic production, distribution, and control of Company documents is allowed and even encouraged (see various Paperwork Reduction Acts, for instance). In all cases, a clear audit trail is required. The question of "Paper or electronic?" is one your Company has to resolve on its own.

Revising And Updating Procedures

As mentioned previously, your policy-and-procedure manual is never complete. It never stops changing or evolving because the needs of the business, its customers' needs, the legal landscape, and the business climate are continually changing.

All employees should be encouraged to initiate changes or revisions to existing policy and procedures that affect their area of responsibility. This greatly assists the Company in keeping the manual current and up-to-date and it gives employees a stake in the Company's well-being.

In addition to this continuing review process, the entire manual should undergo a complete audit as often as changing conditions, both within and outside the company, dictate. This may be every six months, annually, or every other year, depending on the business environment.

A new procedure should be issued if an existing one is to be modified in any way. The revised procedure should undergo the same approval process as the initial release and should be assigned a new revision number level to indicate that it supersedes the prior procedure. Superseded procedures should be purged from the manual immediately and discarded.

[This page intentionally left blank.]

Food Safety Management System Policies, Procedures & Forms

Section 300

Food Safety Manual

Section 300

Food Safety Manual

Food Safety Manual

The Food Safety Manual establishes and states the policies governing the Company's Food Safety Management System (FSMS). These policies define management's arrangements for managing operations and activities in accordance with ISO 22000:2005. These top-level policies represent the plans or protocols for achieving food safety, quality assurance and customer satisfaction.

[This page intentionally left blank]

<Company Logo>

<Our Company, Inc.>

FOOD SAFETY MANUAL

<dd mm yyyy>

Revision #: ______________________________ Effective Date: ___________

Approved by: ______________________________ Date: ___________
President/CEO

The following document contains a sample food safety manual covering the requirements of ISO 22000:2005. This sample is intended only to provide an example of wording that might be used in a food safety manual.

This sample wording can be helpful in generating ideas for developing a manual for your own company. However, food safety policies should be drafted as appropriate and as necessary to accurately reflect your company's food safety management system.

While this manual generally follows the section numbering used in the ISO standard for clarity, it is not necessary for your food safety manual to do the same. The numbering scheme that is most useful and meaningful to you should be selected.

FOOD SAFETY MANUAL

Table of Contents*

1.0 Purpose.......... 1

2.0 Scope.......... 1

3.0 Relation to ISO 22000:2005 1

4.0 Our Company's Food Safety Management System 1
4.1 General Requirements.......... 1
4.2 Documentation requirements.......... 3

5.0 Management Responsibility.......... 4
5.1 Management Commitment.......... 4
5.2 Food Safety Policy.......... 4
5.3 FSMS Planning.......... 5
5.4 Responsibility and Authority.......... 5
5.5 Food Safety Team Leader.......... 5
5.6 Communication.......... 6
5.7 Emergency Preparedness And Response.......... 7
5.8 Management Review 7

6.0 Resource Management.......... 8
6.1 Provision of Resources 8
6.2 Human Resources 9
6.3 Infrastructure.......... 9
6.4 Work Environment.......... 9

7.0 Planning and Realization of Safe Products.......... 9
7.1 General.......... 9
7.2 Prerequisite Programs.......... 10
7.3 Hazard Analysis Preparation.......... 12
7.4 Hazard Analysis.......... 14
7.5 Managing the HACCP Plan.......... 16
7.6 Updating the FSMS.......... 17
7.7 Verification Planning.......... 18
7.8 Traceability.......... 18
7.9 Control of Nonconformity 19

8.0 Validation, Verification, and Improvement of the Food Safety Management System.......... 22
8.1 General.......... 22
8.2 Validation of Control Measure Combinations.......... 22
8.3 Control of Monitoring and Measuring.......... 23
8.4 Food Safety Management System Verification.......... 23
8.5 Continual Improvement.......... 25

FSMS Glossary.......... 27

*The Food Safety Manual Table of Contents references section page numbers (lower right corner).

List of Referenced Procedures

FS1000 – Document Control
FS1010 – Food Safety Records
FS1020 – Management Responsibility
FS1030 – Competence, Awareness, and Training
FS1040 – Job Descriptions
FS1050 – Prerequisite Programs
FS1060 – Hazard Analysis Preparation
FS1070 – Hazard Analysis
FS1080 – HACCP Plan Management
FS1090 – Purchasing
FS1100 – Supplier Evaluation
FS1110 – Receiving and Inspection
FS1120 – Manufacturing
FS1130 – Identification, Labeling, and Traceability
FS1140 – Control of Monitoring and Measuring
FS1150 – Control of Nonconforming Product
FS1160 – Internal Audit and System Validation
FS1170 – Corrective Action
FS1180 – Continuous Improvement
FS1190 – Product Recall
FS1200 – Emergency Preparedness and Response

1.0 PURPOSE

The purpose of this food safety manual is to establish and state the general policies governing Our Company's Food Safety Management System. These policies define management's intended arrangements for managing Company operations and activities in accordance with the framework established by ISO 22000:2005. These are the top-level policies representing the company's plans or protocol for achieving food safety, quality assurance, and customer satisfaction.

All departmental or functional policies and procedures written must conform and parallel these policies. All changes to policies and procedures are required to be reviewed to ensure that there are no conflicts with the policies stated in this Food Safety Manual (FSM).

2.0 SCOPE

The policies stated in this manual apply to all operations and activities at all Company sites (locations). The scope of our food safety system may be stated as in the following example:

"The processing and distribution of seafood products."

It is the responsibility of all department managers to help define, implement, and maintain the procedures required by this manual and to ensure that all internal processes related to food safety conform to these requirements. It is the responsibility of all employees to follow procedures that implement these policies and to help strive for continuous improvement in all activities and processes of Our Company.

EXCLUSIONS

- None

3.0 RELATION TO ISO 22000:2005

For ease of reference, the sections of this manual are *generally* aligned with the equivalent sections of the ISO 22000:2005 standard by section number. The Company may use any numbering system it deems appropriate.

4.0 OUR COMPANY'S FOOD SAFETY MANAGEMENT SYSTEM

4.1 GENERAL REQUIREMENTS

Through this manual and its associated procedures and documents, Our Company has established, documented, implemented, and will maintain a Food Safety Management System conforming to the requirements of ISO 22000:2005. The system is designed to ensure continual improvement of the effectiveness of Our Company in the operation of the Food Safety Management System and in our ability to satisfy our customers' requirements.

Maintenance of this system is the responsibility of the ISO Management Representative in conjunction with all Department Managers.

This Food Safety Manual, along with the associated procedures, identifies the processes needed for the Food Safety Management System at Our Company (Figure 1).

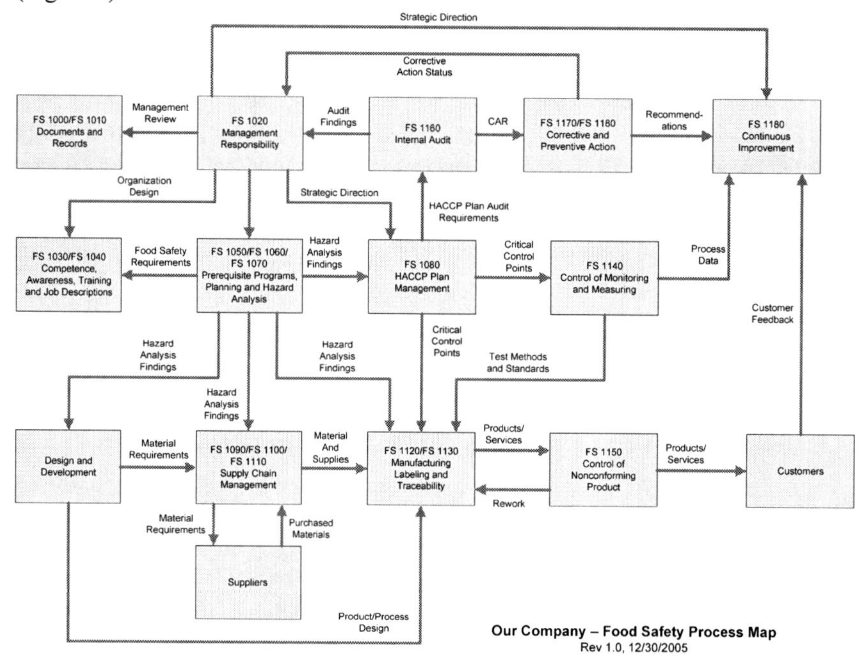

Figure 1 – General Process Sequence Flow Chart

The Food Safety Team Leader maintains a document that identifies the sequence of these processes and, in conjunction with the appropriate department managers, defines the interactions of the processes within the procedures defining these processes. Processes for management activities, provision of resources, product realization, and measurement are included. Procedures shall include the methods needed to ensure the effective operation and control of processes. These processes will be managed in accordance with requirements of ISO 22000:2005.

Top Management will ensure the availability of resources to support the operation and monitoring of processes through regular interaction with department managers and through review activities at Management Review meetings. Department Managers and the Management Rep will monitor, measure, and analyze processes and implement any actions necessary to achieve intended results and continual improvement of the processes. These results will also be monitored at Management Review meetings.

Any outsourced processes that may affect our product's safety and conformity to requirements shall be controlled. The Food Safety Team Leader and appropriate department manager(s) are responsible for defining the methods to control outsourced processes in procedures.

4.2 DOCUMENTATION REQUIREMENTS

4.2.1 General

This Food Safety Manual and the associated procedures are intended to satisfy the ISO 22000:2005 documentation requirements for a food safety manual and food safety objectives. Records required by the ISO 22000 standard are identified in the appropriate procedures and/or the Food Safety Records procedure (FS1010).

Department managers and supervisors are responsible for identifying any additional documents needed to ensure effective planning, operation, and control of processes.

Procedures may vary in detail, based on the size of the department or organization involved, the nature of its business, and the particular type of activity performed. Procedure developers shall consider this, as well as the complexity of the processes, their interactions, and the competence of the personnel involved. Where competence is used to minimize the content in procedures, records must support the decision (see FSM section 6.2.2 – Competence, Awareness, and Training).

Documents may be in any medium, including software programs, electronic files (".doc", ".html", ".pdf", and other formats), and hardcopy documents.

4.2.2 Control of Documents

All documents required by the Food Safety Management System shall be controlled. Controls will ensure that all proposed changes are reviewed prior to implementation to determine their effects on food safety and their impact on the FSMS. The document control procedure will be established to define the controls needed to:

a) Approve documents for adequacy prior to issue;
b) Review and update, as necessary, and approve document revisions;
c) Ensure that changes and the current revision status of documents are identified;
d) Ensure that relevant versions of applicable documents are available at points of use;
e) Ensure that documents remain legible and readily identifiable;
f) Ensure that relevant documents of external origin are identified and their distribution controlled; and
g) Prevent unintended use of obsolete documents and apply suitable identification to them if they are retained for any purpose.

4.2.3 Control of Records

Procedures define appropriate records to be established and maintained in order to provide evidence of conformity to requirements and of the effective operation of the Food Safety Management System. Records shall remain legible, readily identifiable, and retrievable.

The food safety records procedure is established to define the controls needed for the correction, identification, storage, protection, retrieval, retention time, and disposition of records.

4.2.4 Referenced Procedures

- FS1000 – DOCUMENT CONTROL
- FS1010 – FOOD SAFETY RECORDS

5.0 MANAGEMENT RESPONSIBILITY

5.1 MANAGEMENT COMMITMENT

Top Management at Our Company shows its commitment to developing and implementing the Food Safety Management System – and its continued improvement – by showing that food safety is supported by the Company's business objectives. Additionally, management commitment is demonstrated by establishing food safety policy, conducting management review meetings, and providing the resources required to meet our objectives for continually improving the effectiveness of our operations and our Food Safety Management System.

Top Management, consisting of the Company President and all department managers, is chartered with communicating to the entire organization the importance of meeting ISO 22000 requirements, as well as statutory and regulatory requirements and our customers' requirements related to food safety.

5.2 FOOD SAFETY POLICY

Our Company has established and implemented a food safety policy that is appropriate to the organization and meets requirements set forth in ISO 22000:2005. The food safety policy is communicated by Top Management throughout the Company. Department managers and supervisors are responsible for ensuring all employees understand the policy. To ensure our policy remains suitable, appropriate and timely, it shall be reviewed at least annually by Top Management.

The <Our Company> Food Safety Policy:

- It is the policy of Our Company to develop and provide products and services that are safe for consumption, as well as meet or exceed customer requirements and comply with all statutory and regulatory requirements. We accomplish this by adhering to our Food Safety Management System and operational methods that recognize food safety and customer satisfaction as our primary goals.
- We strive to continually improve the effectiveness of our Food Safety Management System and our commitment to customer satisfaction by monitoring our performance against our established objectives and through leadership that promotes employee involvement. This concept represents Our Company's commitment to provide safe food products and to continually improve in order to better serve our growing and demanding customer base.

5.3 FSMS PLANNING

5.3.1 Food Safety Objectives

Top Management shall establish food safety objectives on an annual basis. These objectives shall be measurable and consistent with the Company's food safety policy. Food safety objectives shall also be reviewed at least annually by Top Management.

5.3.2 Food Safety Management System Planning

As part of annual strategic planning meetings, Top Management establishes strategic objectives for improving the Company's products, processes, and customer satisfaction. These objectives are supported by specific measures that track performance against those objectives. Department managers, in turn, set departmental objectives with specific performance measures and targets that support the company objectives.

As situations that demand changes to the Food Safety Management System arise – either to meet objectives or because of changing business conditions – all changes will be reviewed by Top Management, to ensure that the integrity of the Food Safety Management System is maintained.

5.4 RESPONSIBILITY AND AUTHORITY

5.4.1 Responsibility and Authority

Responsibilities and authorities at Our Company are defined in each Job Description, as well as in the Management Responsibility procedure, to ensure effective operation and maintenance of the FSMS. Job Descriptions are also used as a basis for annual performance reviews and are posted on the Company intranet, in addition to being available through Human Resources.

5.4.2 Referenced Procedures

- FS1020 – MANAGEMENT RESPONSIBILITY
- FS1030 – JOB DESCRIPTIONS

5.5 FOOD SAFETY TEAM LEADER

5.5.1 The Company President shall appoint a Food Safety Team Leader. Irrespective of his/her other responsibilities, the Food Safety Team Leader shall have responsibility and authority to:

a) Manage a Food Safety Team (see 7.3.2) and organize its work;
b) Ensure relevant training and education for Food Safety Team members (see 6.2.1);
c) Ensure that the Food Safety Management System is established, implemented, maintained, and updated; and
d) Report to Top Management on the suitability and effectiveness of the FSMS.

5.5.2 The Food Safety Team Leader may additionally be the Company's liaison with external parties on matters relating to the FSMS.

5.5.3 Referenced Procedures

- FS1060 – HAZARD ANALYSIS PREPARATION
- FS1030 – COMPETENCE, AWARENESS, AND TRAINING

5.6 COMMUNICATION

5.6.1 External communication

To ensure that sufficient information on issues concerning food safety is available throughout the food supply chain, the Company shall establish, implement, and maintain effective arrangements for communicating with:

a) Suppliers and contractors;
b) Customers or consumers, particularly with regard to product information (including instructions regarding intended use, specific storage requirements and, as appropriate, shelf life); enquiries; contracts or order handling, including amendments; and customer feedback, including customer complaints;
c) Statutory and regulatory authorities; and
d) Other organizations that have an impact on – or will be affected by – the effectiveness or updating of the Company's Food Safety Management System.

Such communication shall provide information on food safety aspects of the Company's products that may be relevant to other organizations in the food supply chain. This applies especially to known food safety hazards that need to be controlled by other organizations in the food chain. Records of external communications shall be maintained (see 4.2.3).

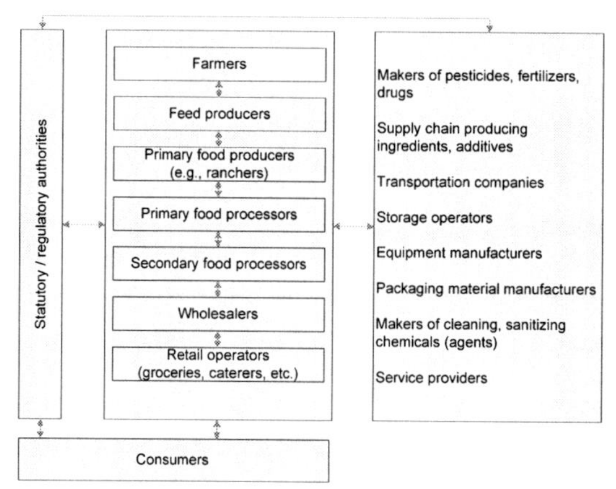

Communicating within the Food Supply Chain (taken from ISO 22000)

Food safety requirements from statutory and regulatory authorities and customers shall be available.

Designated personnel shall have defined responsibility and authority to externally communicate any information concerning food safety. Information obtained through external communication shall be included as input to system updating (see 8.5) and management review (see 5.8).

5.6.2 Internal communication

The Company shall establish, implement, and maintain effective arrangements for communicating with all personnel on issues potentially having an impact on food safety. To maintain the effectiveness of the Food Safety Management System, the Company shall ensure that the Food Safety Team is informed in a timely manner of changes, including – but not necessarily limited to – the following:

a) Existing or new products;
b) Raw materials, ingredients, and services;
c) Production systems and equipment;
d) Production premises, location of equipment, and surrounding environment;
e) Cleaning and sanitation programs;
f) Packaging, storage, and distribution systems;
g) Personnel qualification levels and/or allocation of responsibilities and authorizations;
h) Statutory and regulatory requirements;
i) Knowledge regarding food safety hazards and control measures;
j) Customer, sector and other requirements that the organization observes;
k) Relevant enquiries from external interested parties;
l) Complaints indicating food safety hazards associated with the product; and
m) Other conditions that may have an impact on food safety.

The Food Safety Team shall ensure that this information is included in the updating of the Food Safety Management System (see 8.5). Top Management shall ensure that relevant information is included as input to the Management Review (see 5.8).

5.6.3 Referenced Procedures

- FS1010 – FOOD SAFETY RECORDS
- FS1020 – MANAGEMENT RESPONSIBILITY
- FS1180 – CONTINUOUS IMPROVEMENT

5.7 EMERGENCY PREPAREDNESS AND RESPONSE

Top Management shall establish, implement, and maintain procedures to manage potential emergency situations and accidents that can impact food safety and which are relevant to the role of the Company in the food supply chain.

5.8 MANAGEMENT REVIEW

5.8.1 General

Top Management shall review the Company's Food Safety Management System at least annually (more frequently, if needed) to ensure its continuing suitability, adequacy, and effectiveness. This review shall include assessing opportunities for improvement and the need for change to the Food Safety Management System, including the food safety policy and objectives.

The Food Safety Team Leader shall be responsible for maintaining Management Review records (see 4.2.3).

5.8.2 Review Input

The input to Management Review shall include, but not necessarily be limited to, information on:

a) Follow-up actions from previous Management Reviews;
b) Analysis of results of verification activities (see 8.4);
c) Changing circumstances that can affect food safety (see 5.6.2);
d) Emergency situations, accidents (see 5.7), and product recalls[1] (see 7.9.4);
e) Reviewing results of system updating activities (see 8.4.2);
f) Review of communication activities, including customer feedback (see 5.6.1); and
g) External audits or inspections.

This information shall be presented in a manner that enables Top Management to relate the information to the stated objectives of the Food Safety Management System.

5.8.3 Review Output

The output from the Management Review shall include decisions and actions related to:

a) Assurance of food safety (see 4.1);
b) Improvement of the effectiveness of the Food Safety Management System (see 8.5);
c) Resource needs (see 6.1); and
d) Revisions of the organization's food safety policy and related objectives (see 5.2).

5.8.4 Referenced Procedures

- FS1020 – MANAGEMENT RESPONSIBILITY
- FS1040 – JOB DESCRIPTIONS

6.0 RESOURCE MANAGEMENT

6.1 PROVISION OF RESOURCES

During planning and budgeting processes – and as needed throughout the year – Top Management shall determine and ensure the availability of resources appropriate to establishing, implementing, maintaining, and updating the Food Safety Management System.

[1] The term "recall" also refers to withdrawal.

6.2 HUMAN RESOURCES

6.2.1 General

The Food Safety Team – as well as any other personnel carrying out activities that have an impact on food safety – shall be competent and have appropriate education, training, skills, and experience. Where the assistance of external experts is required for development, implementation, operation, or assessment of the Food Safety Management System, records of agreement or contracts defining the responsibility and authority of external experts shall be available.

6.2.2 Competence, Awareness, and Training

The Company shall:

a) Identify the necessary competencies for personnel whose activities have an impact on food safety;
b) Provide training or take other action to ensure personnel possess the necessary competencies;
c) Ensure that personnel responsible for monitoring, corrections, and corrective actions of the Food Safety Management System are trained;
d) Evaluate the implementation and effectiveness of a), b), and c);
e) Ensure that all personnel are aware of the relevance and importance of their individual activities in contributing to food safety; and
f) Ensure that the requirement for effective communication (see 5.6) is understood by all personnel whose activities have an impact on food safety.

Human Resources shall maintain appropriate records of training and actions described in b) and c).

6.2.3 Referenced Procedures

- FS1030 – COMPETENCE, AWARENESS, AND TRAINING
- FS1040 – JOB DESCRIPTIONS

6.3 INFRASTRUCTURE

Our Company shall provide adequate resources for establishment and maintenance of the infrastructure necessary to implement the requirements of the ISO 22000:2005 standard.

6.4 WORK ENVIRONMENT

The Company shall provide adequate resources for establishment, management, and maintenance of the work environment necessary to implement the requirements of the ISO 22000:2005 standard.

7.0 PLANNING AND REALIZATION OF SAFE PRODUCTS

7.1 GENERAL

The Company shall plan and develop processes needed for the realization (production) of safe products. The Company shall implement, operate, and ensure the effectiveness of planned activities, as well as any changes to those activities.

This includes prerequisite programs (PRPs, or Good Practices – see 7.2), as well as operational PRPs and/or the Company HACCP Plan.

7.2. PREREQUISITE PROGRAMS

7.2.1 Purpose of Prerequisite Programs (PRP)

Top Management will be responsible (or will assign responsibility) for the establishment, implementation, and maintenance of Company PRPs – also known as "Good Practices" or "Standard Operating Procedures" – to assist in controlling:

a) The *likelihood* of introducing food safety hazards to products through the work environment;
b) Biological, chemical, and physical *contamination* of products, including cross-contamination between products; and
c) Food safety hazard levels in the product and product processing *environment.*

7.2.2 PRP Requirements

The PRP shall:

a) Be appropriate to the Company's needs with regard to food safety;
b) Be appropriate to the size and type of the operation and the nature of the products manufactured and/or handled by the Company;
c) Be implemented across the entire production system, either as programs applicable in general or to particular products or operational lines; and
d) Be approved by the Food Safety Team.

The Company's authorized legal representative (or its Legal department) shall identify statutory and regulatory requirements pertaining to prerequisite programs, good practices, and/or SOPs.

7.2.3 Establishing and Maintaining PRPs

When selecting and/or establishing PRPs, Top Management or its representative(s) shall consider and utilize appropriate information (e.g., statutory and regulatory requirements, customer requirements, recognized guidelines, Codex Alimentarius Commission (or Codex) principles and codes of practices[2], and national, international, or sector standards).

The Company shall consider the following when establishing its prerequisite program(s):

a) Construction and layout of buildings and associated utilities;
b) Layout of premises, including workspace and employee facilities;
c) Supplies of air, water, energy, and other utilities;
d) Supporting services, including waste and sewage disposal;
e) Suitability of equipment and its accessibility for cleaning, maintenance, and preventive maintenance;
f) Management of purchased materials (e.g., raw materials, ingredients, chemicals and packaging), supplies (e.g., water, air, steam and ice), disposals

[2] See "annex C" of the ISO 22000:2005 standard, which lists Codex references and guidelines.

(e.g., waste and sewage) and handling of products (e.g., storage and transportation);

g) Measures for the prevention of cross-contamination;
h) Cleaning and sanitizing;
i) Pest control;
j) Personnel hygiene; and
k) Other aspects, as appropriate.

Verification of PRPs shall be planned (see 7.7) and PRPs shall be modified as necessary (see 7.6). Records of verifications and modifications shall be maintained.

Documents should specify how activities included in the Company PRPs are managed.

7.2.4 Establishing Operational Prerequisite Programs

Prerequisite programs are generally well-established and accepted operating guidelines or "Good Practice". PRPs are established *prior to* any hazard analysis (see 7.3 and 7.4). In performing the hazard analysis, a Food Safety Team will identify certain PRPs as ***operational PRPs*** – PRPs ***essential*** to controlling the likelihood of:

- *Introducing* food safety hazards to;
- *Contaminating*; and/or
- *Allowing* food safety hazards to proliferate in

the Company's product(s) or processing environment(s).

The Food Safety Team will document operational PRPs, including the following information in each program:

- Food safety hazard(s) to be controlled by the program (see 7.4.4);
- Control measure(s) (see 7.4.4);
- Modifications, as needed (see 7.7);
- Monitoring procedures that demonstrate that the operational PRPs are in place (see 7.8(a));
- Corrections and corrective actions to be taken if monitoring shows that the operational PRPs are not in control (see 7.9.1 and 7.9.2, respectively);
- Validation of control measures (see 8.2);
- Responsibilities and authorities; and
- Record(s) of monitoring.

7.2.5 Referenced Procedures

- FS1050 – PREREQUISITE PROGRAMS
- FS1150 – CONTROL OF NONCONFORMING PRODUCT
- FS1160 – INTERNAL AUDIT AND SYSTEM VALIDATION
- FS1170 – CORRECTIVE ACTION

7.3 HAZARD ANALYSIS PREPARATION

7.3.1 General

All relevant information needed to conduct hazard analyses shall be collected, maintained, updated, and documented. Records of all hazard analyses shall be maintained.

7.3.2 Food Safety Team

A Food Safety Team will be appointed for each hazard analysis. This Team shall have a combination of multidisciplinary knowledge and experience in developing and implementing the Company's FSMS. Knowledge and experience shall include, but need not be limited to, the Company's products, processes, equipment, and food safety hazards within the scope of the FSMS.

Human Resources will maintain records demonstrating that the Food Safety Team has the requisite knowledge and experience (see 6.2.2).

7.3.3 Product Characteristics

7.3.3.1 Raw materials, ingredients, and product contact materials

All raw materials, ingredients, and product contact materials shall be documented to the extent necessary to conduct hazard analyses (see 7.4). Documentation shall include the following (where appropriate):

a) Biological, chemical and physical characteristics;
b) Composition of formulated ingredients, including additives and processing aids;
c) Origin;
d) Method of production;
e) Packaging and delivery methods;
f) Storage conditions and shelf life;
g) Preparation and/or handling prior to use or processing; and
h) Food-safety-related acceptance criteria or specifications of purchased materials and ingredients appropriate to their intended uses.

The Company shall have appropriate legal counsel identify statutory/regulatory food safety requirements pertinent to the above.

Descriptions of raw materials, ingredients, and product contact materials shall be kept up-to-date (see 7.6).

7.3.3.2 End Product Characteristics

Characteristics of end products shall be documented to the extent necessary to conduct hazard analyses (see 7.4). Documentation shall include information on the following (where appropriate):

a) Product name or similar identification;
b) Composition;
c) Biological, chemical, and physical characteristics relevant to food safety;
d) Intended shelf life and storage conditions;

e) Packaging;
f) Labeling relating to food safety and/or instructions for handling, preparation, and usage; and
g) Method(s) of distribution.

The Company shall have appropriate legal counsel identify statutory/regulatory food safety requirements pertinent to the above.

Descriptions of end products shall be kept up-to-date (see 7.6).

7.3.4 Intended Use

The intended use, reasonably expected handling of the end product, and any unintended (but reasonably expected) mishandling and misuse of the end product shall be considered and documented to the extent necessary to conduct hazard analyses (see 7.4).

Groups of users and – where appropriate – groups of consumers shall be identified for each product and consumer groups known to be especially vulnerable to specific food safety hazards (e.g., food allergies) shall be considered.

Such intended-use descriptions shall be kept up-to-date (see 7.6).

7.3.5 Flow Diagrams, Process Steps, and Control Measures

7.3.5.1 Flow Diagrams

Flow diagrams shall be prepared for products or process categories covered by the FSMS. Flow diagrams provide a basis for evaluating the possible occurrence, increase, or introduction of food safety hazards into the Company's processes and products.

Flow diagrams shall be clear, accurate and sufficiently detailed. Flow diagrams shall include, but need not be limited to, the following (where appropriate):

a) Sequence and interaction of all steps in a given operation;
b) Outsourced processes and subcontracted work;
c) Where raw materials, ingredients, and intermediate products enter the process;
d) Where reworking and recycling take place; and
e) Where end products, intermediate products, by-products, and waste are released or removed.

In accordance with verification planning (see 7.7), the Food Safety Team will verify the accuracy of flow diagrams by process walkthroughs and on-site checking. The Food Safety Team shall maintain verified flow diagrams as records (see 4.2.3).

7.3.5.2 Description of Process Steps and Control Measures

Existing control measures, process parameters (and/or the rigor with which they are applied), or procedures that may influence food safety shall be described to the extent necessary to conduct hazard analyses (see 7.4).

External requirements (e.g. from regulatory authorities or customers) that may impact the choice and the rigorousness of the control measures shall also be described.

Descriptions of process steps and control measures shall be kept up-to-date (see 7.6).

7.3.6 Referenced Procedures

- FS1010 – FOOD SAFETY RECORDS
- FS1030 – COMPETENCE, AWARENESS, AND TRAINING
- FS1060 – HAZARD ANALYSIS PREPARATION
- FS1070 – HAZARD ANALYSIS
- FS1080 – HACCP PLAN MANAGEMENT
- FS1160 – INTERNAL AUDIT AND SYSTEM VALIDATION

7.4 HAZARD ANALYSIS

7.4.1 General

The Food Safety Team will conduct hazard analyses to determine which hazards need to be controlled, the degree of control needed to ensure food safety in each case, and which combination of control measures is required for the hazard(s) in question.

7.4.2 Hazard Identification and Determination of Acceptable Levels

7.4.2.1 The Food Safety Team will identify and record all food safety hazards reasonably expected to occur pertaining to the type of product, type of process, and nature and makeup of processing facilities. Hazard identification will be based on:

a) The preliminary information and data collected, according to 7.3;
b) Experience;
c) External information including, to the extent possible, epidemiological and other historical data; and
d) Information from the food chain on food safety hazards that may be of relevance for the safety of the end products, intermediate products, and the food at consumption.

The Food Safety Team will indicate the step(s) at which each food safety hazard may be introduced.

7.4.2.2 When identifying food safety hazards, the Food Safety Team will consider:

a) The steps preceding and following the specified operation;
b) The process equipment, utilities/services, and surroundings; and
c) The preceding and following links (the Company's suppliers and customers) in the food supply chain.

7.4.2.3 For each food safety hazard identified, the Food Safety Team will, whenever possible, determine an "acceptable level" of each food safety hazard in the end product. In doing so, the FST will take into account established statutory/regulatory requirements, customer food safety requirements, the intended use of the product by the customer (see 7.3.4), and other relevant data. The FST will record the justification for, as well as the result of, the determination.

7.4.3 Hazard Assessment

For each food safety hazard identified by the FST (see 7.4.2), a hazard assessment will be conducted to determine:

a) If eliminating the hazard or reducing it to acceptable levels is essential to the production of safe food; and

b) If control of the hazard is needed to meet the defined acceptable levels.

The FST will evaluate each food safety hazard according to the possible *severity* of adverse health effects and the *likelihood* of its occurrence (see "food safety hazard" and "risk" in the FSMS Glossary). The FST will describe its evaluation methodology for each hazard and record the results of each hazard assessment.

7.4.4 Selection and Assessment of Control Measures

Based on the food safety hazard assessment (7.4.3), the Food Safety Team will identify and select an appropriate combination of control measures, capable of preventing, eliminating, or reducing the food safety hazard(s) to defined acceptable levels.

Each control measure (as described in 7.3.5.2) will be reviewed with respect to its effectiveness against identified food safety hazards.

Each selected control measure will be managed (a) through operational PRP(s) or (b) according to the Company's HACCP Plan, using a logical approach that includes assessments regarding:

a) The control measure's effect on identified food safety hazards, relative to the strictness applied;

b) Feasibility of monitoring (i.e., ability to monitor the control measure in a timely manner, to enable *immediate* corrective action);

c) Its place in the system, relative to other control measures;

d) The likelihood of a control measure failing or exhibiting significant process variability;

e) The severity of the consequences, in the case of control measure failure;

f) Whether the control measure is specifically established and applied to eliminate or significantly reduce the level of food safety hazards; and

g) Synergistic effects (i.e., interaction between control measures that results in their combined effect being greater than the sum of their individual effects).

Control measures categorized as part of the Company HACCP Plan shall be implemented in accordance with 7.5. Other control measures shall be implemented as operational PRPs, in accordance with 7.2. The QA department will document the methodology and parameters used for this categorization and record the results of the assessment.

7.4.5 Referenced Procedures

- FS1060 – HAZARD ANALYSIS PREPARATION
- FS1070 – HAZARD ANALYSIS
- FS1080 – HACCP PLAN MANAGEMENT

7.5 MANAGING THE HACCP PLAN

7.5.1 The HACCP Plan

The HACCP Team will document each HACCP Plan the Company requires for the realization of its product(s). Each HACCP Plan will include the following information for every critical control point (CCP):

a) Food safety hazard(s) to be controlled at the CCP (see 7.4.4);
b) Control measure(s) (see 7.4.4);
c) Critical limit(s) (see 7.5.3);
d) Monitoring procedure(s) (see 7.6.4);
e) Corrections and corrective action(s) to be taken if critical limits are exceeded (see 7.5.5);
f) Responsibilities and authorities; and
g) Record(s) of monitoring.

7.5.2 Identification of Critical Control Points

For each food safety hazard to be controlled by the HACCP plan, the HACCP Team will identify CCPs for identified control measures (see 7.4.4).

7.5.3 Establishment of Critical Limits

The HACCP Team will determine critical limits for the monitoring established for each CCP. Critical limits will be established to ensure that the identified acceptable levels of food safety hazards in the end product (see 7.4.2) are not exceeded. Critical limits shall be measurable. The HACCP Team will document the rationale for each critical limit chosen.

Critical limits based on *subjective* data (e.g., visual inspection) shall be supported by instructions or specifications and/or education and training.

7.5.4 System for Monitoring Critical Control Points

The HACCP Team will establish a monitoring system (if one does not exist) for each CCP to demonstrate that the CCP is in control. The monitoring system will include all scheduled measurements or observations relative to critical limits.

The monitoring system will consist of relevant procedures, instructions, and records that cover the following:

a) Measurements or observations that provide results within an adequate time frame;
b) Monitoring devices used;

c) Applicable calibration methods (see 8.3);
d) Monitoring frequency;
e) Responsibility and authority related to monitoring and evaluation of monitoring results; and
f) Record requirements and methods.

Monitoring methods and frequency must be capable of determining when critical limits have been exceeded in time for the monitored product to be isolated *before* it is used or consumed.

7.5.5 When Monitoring Results Exceed Critical Limits

Planned corrections and corrective actions to be taken when critical limits are exceeded are specified in the Company's HACCP Plan(s). Such actions are designed to ensure that:

a) The cause of nonconformity is identified;

b) The parameter(s) controlled at the CCP is (are) brought back under control; and

c) Recurrence is prevented (see 7.9.2).

Documented procedures shall be established and maintained for the appropriate handling of potentially unsafe products to ensure that they are not released until they have been evaluated (see 7.9.3).

7.5.6 Referenced Procedures

- FS1050 – PREREQUISITE PROGRAMS
- FS1080 – HACCP PLAN MANAGEMENT
- FS1140 – CONTROL OF MONITORING AND MEASURING
- FS1150 – CONTROL OF NONCONFORMING PRODUCT
- FS1170 – CORRECTIVE ACTION

7.6 UPDATING THE FSMS

7.6.1 Updating FSMS Information

Following the establishment of operational PRPs and/or the Company's HACCP plan(s), the Food Safety Team Leader will ensure that the following information is updated, if necessary:

a) Product characteristics;
b) Intended use;
c) Flow diagrams;
d) Process steps; and/or
e) Control measures.

See section 7.3, "Hazard Analysis Preparation", for clarification.

The Food Safety Team Leader will amend the Company HACCP Plan(s) and procedures and instructions specifying prerequisite programs, if needed.

7.6.2 Referenced Procedures

- FS1000 – DOCUMENT CONTROL
- FS1050 – PREREQUISITE PROGRAMS
- FS1060 – HAZARD ANALYSIS PREPARATION
- FS1080 – HACCP PLAN MANAGEMENT

7.7 VERIFICATION PLANNING

7.7.1 Purpose

Verification planning shall define the purpose, methods, frequencies, and responsibilities for verification activities. Verification activities shall confirm that:

a) Prerequisite programs are implemented;
b) Hazard analysis inputs are continually updated;
c) Operational PRPs and elements of the Company HACCP plan(s) are implemented and effective;
d) Hazard levels fall within identified acceptable levels; and
e) Other procedures required by the organization are implemented and effective.

The output of this planning shall be in a form suitable for the Company's method(s) of operations.

7.7.2 Verification Results

Quality Assurance (or a third-party auditor) will record verification results and communicate them to the food safety team. Verification results shall be provided to enable the analysis of the results of the verification activities (see 8.4.3).

7.7.3 Product Control

If system verification is based on testing of end product samples and where such test samples show lack of conformity with the acceptable level(s) of the food safety hazard (see 7.4.2), the affected lots of product shall be handled as potentially unsafe in accordance with 7.9.3.

7.7.4 Referenced Procedures

- FS1070 – HAZARD ANALYSIS
- FS1150 – CONTROL OF NONCONFORMING PRODUCT
- FS1160 – INTERNAL AUDIT AND SYSTEM VALIDATION

7.8 TRACEABILITY

7.8.1 Traceability System

The Company will develop and implement a traceability system that enables the identification of product lots and their relationships to batches of raw materials, processing, and delivery records. This traceability system will enable the

Company to identify incoming material from its immediate suppliers and identify the initial distribution route of the end product.

7.8.2 Traceability Records

Quality Assurance and/or the Food Safety Team Leader will maintain traceability records for a defined period for the purpose of system assessment: (a) to enable the appropriate handling of potentially unsafe products; and (b) in the event of a product recall (withdrawal). Records shall be in accordance with statutory and regulatory requirements and customer requirements and may, for example, be based on the end product lot identification.

7.8.3 Referenced Procedures

- FS1010 – FOOD SAFETY RECORDS
- FS1130 – IDENTIFICATION, LABELING, AND TRACEABILITY
- FS1150 – CONTROL OF NONCONFORMING PRODUCTS

7.9 CONTROL OF NONCONFORMITY

7.9.1 Corrections

The Company will ensure that when critical limits for CCP(s) are exceeded or there is a loss of control of operational PRPs, the affected end products are identified and controlled with regard to their use and release.

A documented procedure shall be established and maintained defining:

a) Identification and assessment of affected end products, to determine their proper handling; and
b) Review of the corrective actions.

Products manufactured under conditions where critical limits have been exceeded are potentially unsafe products and shall be handled in accordance with 7.9.3.

Products manufactured under conditions that do not conform to operational PRPs shall be evaluated with respect to cause(s) of nonconformity and possible food safety consequences and such evaluation will be recorded. Any product made under such conditions will be handled in accordance with 7.9.3, if necessary.

All corrections will be recorded by the party performing such activity and will include information on the nature of the nonconformity, its cause(s) and consequence(s), and information needed for traceability purposes related to the nonconforming lots. Corrections will be approved by Quality Assurance.

7.9.2 Corrective Actions

Data derived from the monitoring of operational PRPs and CCPs shall be evaluated by designated person(s) who possess sufficient knowledge and authority to initiate corrective actions. Corrective actions may be initiated by any employee when he/she observes:

a) Critical limits are being exceeded; or
b) The process is not conforming to operational PRPs.

The Company will establish and maintain documented procedures that specify appropriate actions to:

a) Identify and eliminate the cause of detected nonconformities;
b) Prevent recurrence of those nonconformities; and
c) Bring the process into control after nonconformities are encountered.

Appropriate actions may include, but are not necessarily limited to:

a) Reviewing nonconformities (including customer complaints);
b) Reviewing trends in monitoring results that may indicate development towards loss of control;
c) Determining the cause(s) of nonconformities;
d) Evaluating the need for action to ensure that nonconformities do not recur;
e) Determining and implementing the actions needed;
f) Recording the results of corrective actions taken; and
g) Reviewing corrective actions taken, to ensure their effectiveness.

Corrective actions will be recorded and such records maintained for a period to be determined by statutory/regulatory, Company, and/or customer requirements.

7.9.3 Handling of Potentially Unsafe Products

7.9.3.1 The Company will prevent nonconforming products from entering the food supply chain, *unless* it is possible to ensure that:

a) The food safety hazard in question has been reduced to the defined acceptable levels;
b) The food safety hazard of concern will be reduced to identified acceptable levels prior to releasing the product; or
c) The product meets defined acceptable levels of the food safety hazard of concern, in spite of the nonconformity.

All lots of product(s) that *may have been affected* by a nonconforming situation will be held under the Company's control until after their evaluation.

If products are determined to be unsafe after they have left the Company's control, the Company will notify relevant interested parties and initiate a recall (withdrawal).

The Company will document the controls and related responses and authorization for dealing with potentially unsafe products.

7.9.3.2 Evaluating For Release

Each lot of product affected by a nonconformity will be considered "safe" and released to the customer(s) *only* when one or more of the following conditions applies:

a) Evidence other than that produced by the monitoring system(s) shows control measures have been effective;
b) Evidence exists to show that the *combined* effect of the control measures for that particular product complies with the performance intended (e.g., identified acceptable levels); and/or

c) Results of sampling, analysis, and/or other verification activities show the affected product lot complies with identified acceptable levels for the food safety hazard(s) concerned.

7.9.3.3 Disposing of Nonconforming Product

Following evaluation, if the lot of product is not acceptable for release, it shall be handled by one of the following activities:

a) Reprocessing or further processing within or outside the Company, to ensure that the food safety hazard is eliminated or reduced to acceptable levels; or
b) Destruction of the product and/or disposal as waste.

7.9.4 Recalls (Withdrawals)

7.9.4.1 To enable and facilitate a timely and complete withdrawal of lots of end products identified as unsafe, Top Management will appoint personnel with the authority to initiate withdrawals and will appoint personnel responsible for executing withdrawals.

7.9.4.2 The Company will establish and maintain a documented procedure for:

a) Notifying relevant interested parties (statutory/regulatory authorities, customers, and/or consumers);
b) Handling recalled (withdrawn) products, as well as affected product lots still within the Company's control (i.e., in stock); and
c) The sequence of actions to be taken in the event of a recall.

Withdrawn products shall be secured or held under supervision until they are destroyed, used for purposes other than originally intended, determined to be safe for the same (or other) intended use, or reprocessed in a manner to ensure they become safe.

7.9.4.3 Cause, scope, and results of product recalls will be recorded and reported to Top Management, for input to Management Reviews (see 5.8).

7.9.4.4 The Food Safety Team and Quality Assurance will periodically verify and record the effectiveness of the Company's product recall program by using appropriate methods (e.g., challenge tests, mock withdrawals, or practice withdrawals).

7.9.5 Referenced Procedures

- FS1000 – DOCUMENT CONTROL
- FS1010 – FOOD SAFETY RECORDS
- FS1030 – COMPETENCE, AWARENESS, AND TRAINING
- FS1070 – HAZARD ANALYSIS
- FS1080 – HACCP PLAN MANAGEMENT
- FS1140 – CONTROL OF MONITORING AND MEASURING
- FS1150 – CONTROL OF NONCONFORMING PRODUCTS
- FS1170 – CORRECTIVE ACTION

- FS1190 – PRODUCT RECALL

8.0 VALIDATION, VERIFICATION, AND IMPROVEMENT OF THE FOOD SAFETY MANAGEMENT SYSTEM

8.1 GENERAL

The Food Safety Team will plan and implement the processes needed to validate control measures and/or control measure combinations and to verify and improve the Food Safety Management System.

8.2 VALIDATION OF CONTROL MEASURE COMBINATIONS

8.2.1 Validation

Prior to implementing control measures to be included in operational PRPs and HACCP Plan(s), and after any change (see 8.5), Quality Assurance will verify that:

a) Selected control measures are capable of achieving the intended control of the food safety hazard(s) for which they are designed; and
b) Control measures are effective and are, in combination, capable of ensuring control of identified food safety hazards to enable the Company to make end products that meet defined acceptable levels.

8.2.2 Reevaluation

If the result of a validation shows that one or both of the above elements cannot be confirmed, the control measure and/or combination of measures shall be modified and reevaluated (see 7.4.4).

Modifications may include changes in control measures (i.e., process parameters, rigorousness, and/or their combination) and/or:

a) Change(s) in raw materials;
b) Manufacturing technologies;
c) End product characteristics;
d) Distribution methods; and/or
e) Intended use(s) of end product(s).

8.2.3 Referenced Procedures

- FS1070 – HAZARD ANALYSIS
- FS1140 – CONTROL OF MONITORING AND MEASURING
- FS1160 – INTERNAL AUDIT AND SYSTEM VALIDATION
- FS1170 – CORRECTIVE ACTION
- FS1180 – CONTINUOUS IMPROVEMENT

8.3 CONTROL OF MONITORING AND MEASURING

8.3.1 Monitoring Equipment and Methods

The Company will provide evidence that the specified monitoring and measuring methods and equipment are adequate to ensure the performance of the monitoring and measuring procedures. Where necessary to ensure valid results, measuring equipment and methods used will be:

a) Calibrated or verified at specified intervals, or prior to use, against measurement standards traceable to international or national measurement standards; where no such standards exist, the basis used for calibration or verification shall be recorded;
b) Adjusted and readjusted, as necessary;
c) Identified to enable the calibration status to be determined;
d) Safeguarded from adjustments that would invalidate measurement results; and
e) Protected from damage and deterioration.

Quality Assurance will maintain records of calibration and verification results.

In addition, QA will assess the validity of the previous measurement results when the equipment or process is found not to conform to requirements. If the measuring equipment does not conform, the Company will take action appropriate for the equipment and any product affected.

QA will maintain records of such assessments and resulting actions.

8.3.2 Monitoring Software

When computer software is used to monitor and measure specified requirements, the ability of that software to satisfy the intended application(s) shall be confirmed. Such testing/confirmation will be performed prior to the initial use of such software and software will be retested/reconfirmed as necessary.

8.3.3 Referenced Procedures

- FS1140 – CONTROL OF MONITORING AND MEASURING
- FS1150 – CONTROL OF NONCONFORMING PRODUCT

8.4 FOOD SAFETY MANAGEMENT SYSTEM VERIFICATION

8.4.1 Internal Audit

The Company will conduct a periodic internal audit of its Food Safety Management System, to determine if the FSMS:

a) Conforms to the planned arrangements, to the FSMS requirements established by the Company, and to the requirements of the ISO 22000:2005 standard; and
b) Is being effectively implemented, managed, and updated.

The audit program will be planned to consider the importance of the processes and areas to be audited, as well as any updates resulting from previous audits (see 8.5 and 5.8). Audit criteria, scope, frequency, and methods shall be defined.

Auditors will be selected and audits conducted to ensure the objectivity and impartiality of the audit process. Auditors cannot audit their own work.

Responsibilities and requirements for planning and conducting audits – and for reporting results and maintaining records – will be defined and the audit procedure documented.

Management responsible for areas being audited will ensure that actions to eliminate detected nonconformities and their causes – corrective and preventive actions – are taken without undue delay. Audit follow-up will include verification of corrective/preventive actions taken and reporting of verification results.

8.4.2 Evaluating Individual Verification Results

The Food Safety Team will systematically evaluate individual results of planned verification (see 7.7). If such verification does not demonstrate conformity with the planned arrangements, the Company will take action to achieve the required conformity. Such action shall include, but may not be limited to, review of:

a) Existing procedures and communication channels (see 5.6 and 7.6);
b) Conclusions of hazard analyses (see 7.4), established operational PRPs (see 7.2), and the Company's HACCP Plan(s) (see 7.5);
c) Prerequisite programs (see 7.2); and
d) The effectiveness of Human Resource management and of training activities (see 6.2).

8.4.3 Analyzing Results of Verification Activities

The Food Safety Team will analyze results of verification activities, including results of the internal audits (see 8.4.1) and external audits, in order to:

a) Confirm that the overall performance of the system meets the planned arrangements and the Food Safety Management System requirements established by the Company;
b) Identify the need for updating or improving the Company's Food Safety Management System;
c) Identify trends that may indicate a higher incidence of potentially unsafe products;
d) Establish information for planning of the internal audit program concerning the status and importance of areas to be audited; and
e) Provide evidence that any corrections and corrective actions taken are effective.

The Food Safety Team will ensure that results of the analysis and resulting activities are recorded and reported in an appropriate manner to Top Management, as input to the Management Review (see 5.8) and that such results will be used as input for updating the FSMS (see 8.5).

8.4.4 Referenced Procedures

- FS1010 – FOOD SAFETY RECORDS
- FS1080 – HACCP PLAN MANAGEMENT

- FS1140 – CONTROL OF MONITORING AND MEASURING
- FS1150 – CONTROL OF NONCONFORMING PRODUCT
- FS1160 – INTERNAL AUDIT AND SYSTEM VALIDATION

8.5 CONTINUAL IMPROVEMENT

8.5.1 Continual Improvement

Top Management shall ensure that the Company continually improves the effectiveness of its FSMS through the use of:

a) Effective communication (see 5.6);
b) Management review (see 5.8);
c) Internal audits (see 8.4);
d) Evaluating individual verification results (see 8.4);
e) Analyzing results of verification activities (see 8.4);
f) Validating control measures or their combinations (see 8.2);
g) Corrective actions (see 7.9); and
h) FSMS updating (see 8.5.2).

8.5.2 Updating the FSMS

Top Management shall ensure that the Company FSMS is continually updated. To achieve this, the Food Safety Team shall periodically evaluate the FSMS. After evaluating the FSMS, the FST will consider the need to review hazard analyses (see 7.4), established operational PRPs (see 7.2), and the Company's HACCP Plans (see 7.5). Evaluation and update activities will be based on:

a) Input from internal and external communications (see 5.6);
b) Other information concerning the suitability, adequacy, and effectiveness of the Food Safety Management System;
c) Output from analyzing results of verification activities (see 8.4); and
d) Output from Management Reviews (see 5.8).

FSMS updating activities shall be recorded and reported in an appropriate manner as input to Management Reviews (see 5.8).

8.5.3 Referenced Procedures

- FS1020 – MANAGEMENT RESPONSIBILITY
- FS1170 – CORRECTIVE ACTION
- FS1180 – CONTINUOUS IMPROVEMENT

Food Safety Manual - Revision History

Revision	Date	Description of changes	Requested By
0.0		Initial Release	

FSMS GLOSSARY

Term	Definition
Acceptable level	The level of a safety hazard considered to present a risk the consumer would accept. The acceptable level of the hazard in the end product, sometimes referred to as the "target level", should be stated in the product description and set at or below statutory/regulatory limits. An acceptable level for a hazard at an intermediate step in the commodity (product) flow diagram may be set higher than that of the final product, provided that the acceptable level in the final product is achieved.
Active record	Record currently in use or used in the context of ongoing business. May also be referred to as a "production" record.
Calibration	Comparison of a measurement standard or instrument of known accuracy with another standard or instrument to detect, correlate, report, or eliminate by adjustment any variation in the accuracy of the item being compared.
Calibration period	Period during which a certified calibration is valid.
CCP	See "critical control point".
CCP decision tree	A sequence of questions to assist in determining whether a control point is a CCP.
Competence	State of having a combination of adequate training and experience to perform a task or set of tasks.
Control measure	An action or activity that can be used to prevent or eliminate a food safety hazard (3.3) or reduce it to an acceptable level.
Control point	Any step at which biological, chemical, or physical factors can be controlled.
Controlled document	Document that provides information or direction for performance of work within the scope of a given procedure. Control characteristics may include, but are not limited to, revision number/letter, revision date, signatures indicating review and approval, and controlled distribution.
Correction	Action taken to eliminate a detected nonconformity. 1) For the purposes of ISO 22000, a correction relates to the handling of potentially unsafe products and can, therefore, be made in conjunction with a corrective action. 2) A correction may be, for example, reprocessing, further processing, and/or elimination of the adverse consequences of the nonconformity (such as disposal for other use or specific labeling).

Term	Definition
Corrective action	Action taken to eliminate the cause of a detected nonconformity or other undesirable situation 1) A nonconformity may have more than one cause. 2) Corrective action includes cause analysis and is taken to prevent recurrence. See "preventive action".
Critical control point (CCP)	A step at which control can be applied that is essential to prevention or elimination of a food safety hazard or reduction of the hazard to an acceptable level.
Critical limit	Criterion that separates the acceptable from the unacceptable. Critical limits are established to determine whether a CCP remains in control. If a critical limit is exceeded or violated, the affected products are deemed potentially unsafe.
Cross-contamination	The transfer of harmful bacteria from one food to another by way of a nonfood surface, such as a cutting board, countertop, utensils, or a person's hands.
Deviation	Failure to meet a critical limit.
Document	Information and its supporting medium. The medium may be paper, magnetic, electronic, optical computer disc, photograph, or sample.
End product	A product that will undergo no additional processing or transformation within the organization. A product that undergoes further processing or transformation by another organization is an end product within the context of the first organization and a raw material or ingredient in the context of the second organization.
External document	A document of external origin that provides information or direction for the performance of activities within the scope of the Food Safety Management System. Examples include, but are not limited to, customer drawings, industry standards, international standards, and equipment manuals.
Farm-to-Table Continuum	A multi-step journey that food travels before it is consumed. The steps in the continuum are Farm, Processing, Transportation, Retail, and Table. Each sector along the farm-to-table continuum plays a role in ensuring that the food supply is fresh, of high quality, and safe from hazards. If a link in the continuum is broken, the safety and integrity of the food supply can be threatened.
Flow diagram	A schematic, systematic presentation of the sequence and interactions of steps in a process. A flow diagram usually takes the form of a flowchart, where all steps in a process and their inputs and outputs (including byproducts and waste) are shown as boxes connected by unidirectional arrows. Flow diagrams may be referred to as "process maps".

Term	Definition
Food Code (USA)	A 400-page reference guide published by the U.S. Food and Drug Administration (FDA). The Food Code instructs retail outlets (such as restaurants and grocery stores) and institutions (such as nursing homes and schools) on how to prevent foodborne illness. It consists of model requirements for safeguarding public health and ensuring that food is unadulterated (free from impurities) and honestly presented to the consumer. The FDA first published the *Food Code* in 1993 and revised it every two years through 2001; at that time, it was agreed that the Food Code would be revised every four years. The last revision was in 2005.
Food safety	The concept that food will not cause harm to the consumer when prepared and/or eaten according to its intended use.
Food safety hazard	Biological, chemical, or physical agent in food or condition of food with the potential to cause an adverse health effect. 1) The term "hazard" is not to be confused with the term "risk" which, in the context of food safety, means a function of the probability of an adverse health effect (e.g. becoming diseased) and the severity of that effect (death, hospitalization, absence from work, etc.) when exposed to a specified hazard. Risk is defined in ISO/IEC Guide 51 as the combination of the probability of occurrence of harm and the severity of that harm. 2) Food safety hazards include allergens. 3) In the context of feed and feed ingredients, relevant food safety hazards are those that may be present in and/or on feed and feed ingredients and that may subsequently be transferred to food through animal consumption of feed and may thus have the potential to cause an adverse human health effect. In the context of operations other than those directly handling feed and food (e.g. producers of packaging materials, cleaning agents, etc.), relevant food safety hazards are those hazards that can be directly or indirectly transferred to food because of the intended use of the provided products and/or services and thus can have the potential to cause an adverse human health effect.
Food Safety Management System (FSMS)	An ordered, well-documented system that results in safe food. The FSMS is designed to ensure consistency and improvement of work procedures and practices, including produced goods. These procedures are based on standards or principles, such as ISO 22000 or HACCP, that specify procedures for achieving effective management in the safety of food production.
Food safety policy	Overall intentions and direction of an organization related to food safety, as formally expressed by top management
Food safety team	Personnel responsible for testing, inspecting, and reporting on FSMS procedures to ensure their conformance to applicable requirements.

Term	Definition
Food safety team leader	Someone who has acquired the necessary competencies, training, certifications, and managerial skills to lead a Food Safety Team.
Food supply chain (or food chain)	A sequence of stages and operations involved in the production, processing, distribution, storage, and handling of food and/or its ingredients, from primary production to consumption. 1) This includes production of feed for food-producing animals and for animals intended for food production. 2) The food (supply) chain also includes the production of materials intended to come into contact with food or raw materials.)
Good practice	A practice or set of practices designed to ensure that food products, services, etc., are executed according to prescribed food safety standards. Good Practice ensures that finished products have the correct identity, strength, quality and purity characteristics they are represented to have, and have not been altered during processing, packaging, or handling. Most "good practices" have been around for so long and are commonly followed by good producers, etc., that standards and regulations have grown up around them. Examples of "good practices" include Good Manufacturing Practice, Good Veterinary Practice, and Good Hygienic Practice.
HACCP	A systematic approach to the identification, evaluation, and control of food safety hazards.
HACCP plan	The written document, based upon the principles of HACCP, which delineates the food safety procedures to be followed by the Company.
HACCP system	The result of implementing the HACCP Plan.
HACCP team	People responsible for developing, implementing, and maintaining the HACCP system.
Hazard analysis	The process of collecting and evaluating information on hazards associated with the food under consideration to decide which are significant and must be addressed in the HACCP plan. Hazard analysis consists of two steps, identification and evaluation.
High-risk food	Food that supports the growth of bacteria and/or microbes, such as meat, dairy, or eggs.
Hold	Time period used for investigation after a food has been identified as potentially unsafe. The "hold" process is unique to USDA commodity foods.
Internal document	Document of *internal* origin (developed entirely by or completed by the Company) that provides information or direction for the performance of activities within the scope of the Food Safety Management System. Examples include, but are not limited to, the procedures contained in the Company's FSMS manual.

Term	Definition
Management team	Consists of the Food Safety Team Leader, Department Managers, and the Company President, at a minimum.
MAP	See "modified atmosphere packaging".
Material Safety Data Sheet (MSDS)	A Material Safety Data Sheet is designed to provide workers and emergency personnel with the proper procedure(s) for handling or working with a particular substance. MSDSs include information such as physical data (melting point, boiling point, flash point, etc.), toxicity, health effects, first aid, reactivity, storage, disposal, protective equipment, and spill/leak procedures that are of particular use if a spill or other accident occurs. An MSDS is designed for employees who may be occupationally exposed to hazards at work, employers who need to know the proper methods for storage, etc., and emergency responders (such as fire fighters, hazardous material (HazMat) crews, emergency medical technicians, and hospital emergency room personnel). MSDSs are not designed for consumers – they reflect the hazards of working with materials occupationally. For example, an MSDS for paint does not apply to someone who uses a can of paint once a year but does apply to someone who uses paint, especially in confined spaces, 40 hours a week.
Modified atmosphere packaging (MAP)	Food packaging in which a mixture of gases replaces ordinary air in the food package. Carbon dioxide and nitrogen are commonly used in MAP to replace oxygen. Many foodborne pathogens cannot thrive in low-oxygen environments. A low-oxygen environment also inhibits spoilage by preventing growth of molds and yeasts.
Monitoring	Conducting a planned sequence of observations or measurements to assess whether control measures are operating as intended; also, the regular measurement or observation of a critical control point to make sure the product does not go outside of its critical limits.
MSDS	See "Material Safety Data Sheet".
Operational prerequisite program	A PRP identified during a hazard analysis as *essential to controlling*: (a) the likelihood of introducing food safety hazards to; (b) contamination of; and/or (c) proliferation of food safety hazards in the product(s) or processing environment(s). Also known as an "operational PRP".
Prerequisite program	Basic conditions and activities necessary to maintain a hygienic environment throughout the food supply chain which is suitable for production, handling, and provision of safe end products and safe food for human consumption. PRPs depend on the segment of the food chain in which the organization operates and the type of organization (see ISO 22000:2005, Annex C). Examples of equivalent terms are Good Agricultural Practice (GAP), Good Veterinarian Practice (GVP), Good Manufacturing Practice (GMP), Good Hygienic Practice (GHP), Good

Term	Definition
	Production Practice (GPP), Good Distribution Practice (GDP), and Good Trading Practice (GTP).
Preventive action	Long term cost / risk weighted action taken to prevent a problem from occurring, based on an understanding of the product or process. See "corrective action".
PRP	Prerequisite program.
Product realization	The act of bring a product (goods or services) into existence; making a product.
Recall	Remove a food product from the market because it may cause health problems or possible death; withdraw.
Reference standard	A standard of the highest order of accuracy in a calibration system, establishing the basic accuracy values for that system. See "working standard".
Risk	A function of the *probability* of an adverse health effect (e.g., disease, illness) and the *severity* of that effect (e.g., work absence, hospitalization, death) when exposed to a specified hazard.
Safe food	Food that is not harmful or injurious when consumed; food that does not cause medical illness or pose a health hazard to the consumer. Recently, food scientists, nutritionists, and various organizations have pushed for a narrower definition of safe food, to include only foods that provide a long-term nutritional benefit or promote health. It is unlikely the narrower definition will ever have full force of law, considering the economic impact it would have on producers, not to mention the success of similar legislation (e.g., Prohibition, 1919-1933, USA). Regardless, one must be mindful of the context in which the term "safe food" is used.
Segregation	Removal of product to an area of storage that spatially (physically) isolates it from other foods.
Supplier	Company/organization that directly supplies Our Company with food; food ingredients; food processing, handling, and/or packaging equipment; and/or other items directly or indirectly related to food safety (e.g., cleaning/sanitation chemicals, labels, containers, equipment maintenance services).
Target	Standard which must be met to control a hazard.
Target level	See "acceptable level".
Traceability	The ability to relate individual measurement results to national standards or nationally accepted measurement systems through an unbroken chain of comparisons.

Term	Definition
Uncontrolled document	Document that was removed from – or never was a part of – the Company's controlled document system. Uncontrolled documents may not be used to provide work direction or information necessary for the performance of work. Uncontrolled copies of documents may be used as training aids.
Updating	Immediate and/or planned activity to ensure application of the most recent information on a given topic.
Validation	Obtaining evidence that the control measures managed by the HACCP plan and by the operational PRPs are capable of being effective
Vendor	See "supplier".
Verification	Confirmation, by obtaining objective evidence, that specified requirements have been fulfilled.
Withdraw	See "recall".
Working standard	Designated measuring equipment used in a calibration system as a medium for transferring the basic value of reference standards to lower echelon transfer standards or other measuring and test equipment. See "reference standard".

[This page intentionally left blank.]

Food Safety Management System
Policies, Procedures & Forms

Section 400

Food Safety Procedures

FS1000 Document Control
FS1010 Food Safety Records
FS1020 Management Responsibility
FS1030 Competence Awareness and Training
FS1040 Job Descriptions
FS1050 Prerequisite Programs
FS1060 Hazard Analysis Preparation
FS1070 Hazard Analysis
FS1080 HACCP Plan Management
FS1090 Purchasing
FS1100 Supplier Evaluation
FS1110 Receiving and Inspection
FS1120 Manufacturing
FS1130 Identification Labeling and Traceability
FS1140 Control of Monitoring and Measuring
FS1150 Control of Potentially Unsafe Product
FS1160 Internal Audit and System Validation
FS1170 Corrective Action
FS1180 Continual Improvement
FS1190 Product Recall
FS1200 Emergency Preparedness and Response

Section 400

Food Safety Procedures

Document #: **FS1000**	Title: **DOCUMENT CONTROL**	Print Date: **2/1/2006**
Rev #: **0.0**	Prepared By:	Date Prepared: **2/1/2006**
Effective Date: **11/18/2005**	Reviewed By:	Date Reviewed: **2/1/2006**
	Approved By:	Date Approved: **2/1/2006**
Standard: **ISO 22000:2005, clause 4.2.2**		

Policy: To control Food Safety Management System documents, ensuring that everyone responsible for food safety is working according to the same set of procedures and guidelines.

Purpose: To define the methods and responsibilities for controlling documents used to provide work direction or set policy; and to define methods for document revision, approval, and distribution.

Scope: This procedure applies to all documents required by the Food Safety Management System. Documents of internal or external origin are included.

Responsibilities:

Document Control is responsible for controlling the Food Safety Management System Manual, all procedures and instructions related to the Food Safety Management System, and all external documents that are required by the Company's FSMS.

Department Managers and supervisors are responsible for ensuring the relevant versions of documents are available at the points of use and that they are legible. They may also be responsible for reviewing and responding to change requests in a timely manner.

Engineering is responsible for control of drawings (external and internal) and bills of material.

The Food Safety Team is responsible for reviewing all procedures and instructions at least annually to ensure documents remain current.

The Food Safety Team Leader is responsible for reviewing requests to change FSMS-related procedures and work instructions and directing the Food Safety Team through such reviews.

Definitions: Controlled Document – Document that provides information or direction for performance of work within the scope of a given procedure. Characteristics of control include such things as revision number (letter), signatures indicating review and approval, and controlled distribution.

Document - Information and its supporting medium. The medium may be paper, magnetic, electronic, optical computer disc, photograph, or sample.

External Document – Document originating outside the Company (e.g., customer drawings, industry and/or international standards, vendors' equipment maintenance manuals) that provides information or direction for the performance of activities within the scope of the Food Safety Management System.

Food Safety Management System (FSMS) – Ordered, well-documented system designed to result in safe food.

Internal Document – Document of *internal* origin (developed entirely by or completed by the Company) that provides information or direction for the performance of activities within the scope of the Food Safety Management System. Examples include, but are not limited to, the procedures contained in this FSMS manual.

Uncontrolled Document – Document removed from, or never a part of, the controlled document system. Uncontrolled documents may not be used to provide work direction or information necessary for the performance of work. Uncontrolled copies of documents may be used as training aids.

Procedure:

1.0 PROCEDURE AND WORK INSTRUCTION FORMAT

1.1 Procedures and instruction should use this document as a template.

1.2 If any of the headings are not applicable they may be deleted.

1.3 All procedures and instructions must have the Procedure Name, Revision and Page Number in Page X of Y format on each page of the document.

1.4 All procedures and instructions must show approval signatures on the first page or cover page. Electronic versions may show the approving parties' names typed but in any case, Document Control shall keep a signed *paper* "master" copy of each document.

1.5 Electronic files shall be named using the document number, the title, and the revision (e.g., "FS1000 – Document Control – rev 0.2")

2.0 TEMPORARY CHANGES TO DOCUMENTS

2.1 For many reasons, temporary changes to the Food Safety Management System may be needed. Temporary changes must be defined and controlled.

2.2 Temporary changes shall be brought to the attention of affected Department Management.

2.3 If a document (e.g., for a procedure) does not exist, Department Management may develop a temporary document, which must bear the Manager's name, title, and current date. If an existing document (e.g., procedure) needs changes, Department Management shall red-line, initial, and date the changes on the existing document. Whether the document in question is new or altered, it must be legible and identifiable.

2.4 Department Management shall submit changes to the Food Safety Team Leader, who shall review, initial, and date the changed document.

2.5 If temporary changes bypass any part of the HACCP Plan, the changes require the approval of Top Management, who shall also review, initial, and date the document.

2.6 Temporary change documents must include the range of dates in which the changes are valid and the scope of applicability. Temporary documents may not be used for more than two weeks without being made permanent.

3.0 DOCUMENT REVISION

3.1 The Food Safety Team Leader is responsible for coordinating with Department Managers to review all procedures and instructions at least annually and update them, as required, to ensure documents remain current.

3.2 Anyone may submit a new document or changes to an existing document, as necessary. To submit a change, the requestor shall print a copy of the document and mark ("redline") the requested changes on the copy. If changes are extensive, a new document may be typed and submitted with a copy of the original. The requestor shall clearly identify, initial, and date all such changes.

3.3 The requestor shall submit changes to his/her Department Manager for review. The Department Manager, after reviewing the requested change(s), shall complete a FS1000-1 – REQUEST FOR DOCUMENT CHANGE (RDC), indicating the nature and reason for the change. He/she shall submit the FS1000-1 with the changed document to the Food Safety Team Leader for review.

3.4 The Food Safety Team Leader shall review the change request to ensure ISO 22000 compliance.

- If the request for change is not in compliance with the standard, the Food Safety Team Leader shall notify the requestor of the reason for the denial and explain the changes needed to bring the request into compliance.
- If the request is ISO 22000 compliant, the Food Safety Team Leader shall determine who else needs to review the change and submit the change and the FS1000-1 form to those individuals (the document review team) for their review.

3.5 The document reviewers shall consider the change and the reason for it and determine if the change is warranted.

- If the document review team approves the requested changes, the Food Safety Team Leader shall assign a Document Change Number (DCN) on FS1000-2 – DOCUMENT CHANGE CONTROL and submit the new or changed documents, with the appropriate approvals and the FS1000-1 form, to Document Control for typing and formatting. Document Control shall update the document, index the revision, update the document status on FS1000-2, and update the revision history on the document.

3.6 Document Control shall notify the document review team via e-mail when the revised document is available for a "final" review. Reviewers shall indicate intended approval or submit comments via e-mail. If comments are substantive, Document Control shall incorporate the comments and contact the document review team for another review.

3.7 When all document review team members indicate their intent to approve, Document Control shall circulate the final document(s) to obtain signatures. When the required signatures have been obtained, Document Control shall update FS1000-2 with the new revision number and revision date for *changed* documents, or with all required information for *new* documents.

3.8 Document Control shall make sufficient copies for distribution to all locations indicated on FS1000-3 – DOCUMENT CONTROL DATABASE. Document Control shall stamp "Controlled Copy" in red on each paper copy before distributing.

- Document Control shall pull the master copy of the previous revision, mark it "Obsolete", and file it in the historical files (archives).
- Because the electronic version will not show signatures, the current Master Copy is maintained as evidence of review and approval.

4.0 DOCUMENT DISTRIBUTION

4.1 Flow diagrams, engineering drawings, and other technical documents are distributed by the responsible department, either electronically or via hard copy. Obsolete drawings and other technical documents are to be returned to the responsible department when revisions are distributed.

4.2 Document Control shall maintain a database of all controlled documents (FS1000-3 – DOCUMENT CONTROL DATABASE). Documents of internal origin and external origin (e.g., customer drawings, national or international standards that may be used or referenced) shall be maintained in separate lists or databases.

The document control database should contain, at a minimum, the following information on each document:

- Document Number;
- Document Title;
- Current Revision number or ID;
- Revision Date;
- Responsible Department; and
- Document Location

4.3 Document Control shall distribute hardcopy documents only to the locations shown on the database and shall remove and destroy any previous versions of procedures or work instructions.

- Document Control shall distribute electronic versions of documents by moving old revisions to the OBSOLETE folder, moving the new revisions to the RELEASED folder, and notifying affected parties in a timely manner.
- All electronic documents moved to the OBSOLETE folder shall be assigned expiration dates. When a document's expiration date is reached, it shall be permanently removed from electronic storage.

4.4 External documents are controlled only for distribution. All external standards should be purchased through Document Control, to ensure they are added to the External Document list and are properly controlled and distributed.

5.0 DOCUMENT CONTROL PROCESS REVIEW

The Food Safety Team Leader shall review the document control process on a regular basis (annually, at a minimum) with Document Control, to ensure that all food-safety-related documents are being controlled effectively and that the process continues to meet Company, statutory/regulatory, and standards-based requirements.

Effectiveness Criteria:

- No obsolete documents in use
- Average time to release document changes

References:

A. Food Safety Manual
   - FSM 4.2.2 – Control of Documents

Records:

- Signed master copies of documents.
- FS1000-1 – REQUEST FOR DOCUMENT CHANGE (RDC)
- FS1000-2 – DOCUMENT CHANGE CONTROL
- FS1000-3 – DOCUMENT CONTROL DATABASE

Revision History:

Revision	Date	Description of changes	Requested By
0.0	2/1/2006	Initial Release	

FS1000-1 – REQUEST FOR DOCUMENT CHANGE (RDC)

Date:______________________________ RDC No.:________________

Originator: ____

Document title and publication date: _____________________________________

Page / chapter / paragraph number: ______________________________________

Description of problem, opportunity or reason for request (define in detail):

____

____

____

Solution recommended (if known) date action required by:____________________

____

____

____

Comments:____

____

____

____

____

Department manager approval:____

Recommended solution to problem or postponement/dissolution of request
(attach all necessary documentation to support response)_____________________

____

____

____

Approved By:____________________________ Date:________________

PROCEDURE FOR COMPLETING FORM

1) Complete top section of this form except for RDC number
2) Obtain Department Manager's approval
3) Forward original to the Food Safety Team Leader who will assign RDC number (Note: one copy will be returned to originator with RDC number assigned. If the related process owner is not the originator, the request will be forwarded to the process owner for action.
4) The process owner will take action and if appropriate will proceed with an RDC.
5) The Food Safety Team Leader returns a copy to Originator upon resolution of request.

Distribution: Original - RDC File Copy 1 - Originator

[This page intentionally left blank]

FS1000-2 – DOCUMENT CHANGE CONTROL

Date:________________

DCN#:_______________

RDC#:_______________

PART I

Doc. or Part No.	Description of Change, Documents affected and reason(s) for change(s)	Action Code(s)	Effective Date

Change Action Required

Make/Order New Document: ______________________________

Current Docs: ❒ Use until depleted ❒ Return for credit ❒ Scrap ❒ Save for spares

Rework Docs: ❒ In stock ❒ In process ❒ Finished goods

Service Repair: ❒ Reference only ❒ Field doc change ❒ Notify customer

Other:____

Comments:______________________________________

PART II

To assure an orderly, controlled and complete implementation of a change to all aspects of related documentation, the following checklist should be completed along with appropriate revisions and updating of all copies of documents. Attach additional paper for instructions or comments if necessary.

Document Item	Affected Yes/No	Description	Assigned To	Date Finished
Engineering/Manufacturing				
Specifications Sheet	❐ ❐	______	______	______
Bill of Materials	❐ ❐	______	______	______
Manufacturing Procedures (Preparation Manuals)	❐ ❐	______	______	______
Testing Procedures	❐ ❐	______	______	______
Inspection / QC Procedures	❐ ❐	______	______	______
Procurement Documentation	❐ ❐	______	______	______
Labeling	❐ ❐	______	______	______
Packaging	❐ ❐	______	______	______
Service Manuals	❐ ❐	______	______	______
Drawings	❐ ❐	______	______	______
______	❐ ❐	______	______	______
______	❐ ❐	______	______	______
Sales/Marketing				
Product Literature	❐ ❐	______	______	______
Price Catalog	❐ ❐	______	______	______
______	❐ ❐	______	______	______
______	❐ ❐	______	______	______
Regulatory Compliance				
FDA	❐ ❐	______	______	______
Government Contracts	❐ ❐	______	______	______
______	❐ ❐	______	______	______
______	❐ ❐	______	______	______

Authorization(s): Engineering

By:______________________

Title:______________________

Date:______________________

Authorization(s): Food Safety

By:______________________

Title:______________________

Date:______________________

FS1000-3 – DOCUMENT CONTROL DATABASE

Document Number	Document Title	Rev Num	Rev Date	Responsible Department	Document Location
FS1000	Document Control	0.0	12/30/05	Document Control	Document server
FS1010	Food Safety Records	0.0	12/30/05	Food Safety	Document server
FS1020	Management Responsibility	0.0	12/30/05	Management Team	Document server
FS1030	Competence, Awareness and Training	0.0	12/30/05	Human Resources	Document server
FS1040	Job Descriptions	0.0	12/30/05	Human Resources	Document server
FS1050	Prerequisite Programs	0.0	12/30/05	Food Safety	Document server
FS1060	Hazard Analysis Preparation	0.0	12/30/05	Food Safety	Document server
FS1070	Hazard Analysis	0.0	12/30/05	Food Safety	Document server
FS1080	HACCP Plan Management	0.0	12/30/05	Food Safety	Document server
FS1090	Purchasing	0.0	12/30/05	Purchasing	Document server
FS1100	Supplier Evaluation	0.0	12/30/05	Purchasing	Document server
FS1110	Receiving and Inspection	0.0	12/30/05	Manufacturing	Document server

Document Number	Document Title	Rev Num	Rev Date	Responsible Department	Document Location
FS1120	Manufacturing	0.0	12/30/05	Manufacturing	Document server
FS1130	Identification, Labeling, and Traceability	0.0	12/30/05	Manufacturing	Document server
FS1140	Control of Monitoring and Measuring	0.0	12/30/05	Quality Assurance	Document server
FS1150	Control of Nonconforming Product	0.0	12/30/05	Quality Assurance	Document server
FS1160	Internal Audit and System Validation	0.0	12/30/05	Food Safety	Document server
FS1170	Corrective Action	0.2	12/30/05	Food Safety	Document server
FS1180	Continuous Improvement	0.0	12/30/05	Food Safety	Document server
FS1190	Product Recall	0.0	12/30/05	Food Safety	Document server
FS1200	Emergency Preparedness and Response	0.0	12/30/05	Food Safety	Document server
FSMS 100	Food Safety Manual	0.0	12/30/05	Food Safety	Document server

Document #: **FS1010**	Title: **FOOD SAFETY RECORDS**	Print Date: **2/1/2006**
Rev #: **0.0**	Prepared By:	Date Prepared: **2/1/2006**
Effective Date: **2/1/2006**	Reviewed By:	Date Reviewed: **2/1/2006**
	Approved By:	Date Approved: **2/1/2006**
Standards: **ISO 22000:2005, clause 4.2.3**		

Policy: To assure our customers of the effectiveness of our FSMS, we will maintain a thorough record of all Company activities, especially with regard to food safety.

Purpose: To demonstrate conformance to specified requirements and ensure effective operation of the Food Safety Management System.

Scope: This procedure applies to all records generated - handwritten, hardcopy, and electronic - that serve to record Our Company's activities related to food safety and are required to demonstrate implementation of and conformance to Our Company's Food Safety Management System.

Responsibilities:

Department Managers are responsible for ensuring that food safety records are accurate, complete, and up-to-date, according to this procedure.

All personnel are responsible for ensuring records they generate are accurate, timely, and legible.

Definitions: Record – Anything, such as a document or a photograph, providing permanent evidence of or information about past events.

Procedure:

1.0 IDENTIFICATION OF FOOD SAFETY RECORDS

1.1 Each department and/or functional group is responsible for maintaining adequate records to demonstrate effective system operations as defined specifically in each procedure and to demonstrate conformance to legal/regulatory and customer requirements.

1.2 FS1010-1 – FOOD SAFETY RECORDS LIST should provide a complete list of the Company's food safety records (the example provided includes a number of records required by the ISO 22000 standard, shown *in bold italics*). Each record listed is described by its title, file location, authority or responsible department, minimum retention time, and directions for disposition.

2.0 FOOD SAFETY RECORD GENERATION

2.1 All personnel may generate records having an effect on food safety while performing their duties. All food safety records must be accurate, timely, legible, and readily accessible.

2.2 Written food safety records should be completed in ink to help ensure legibility and to protect them from unauthorized change.

2.3 Changes or corrections to hardcopy records should be made with a single line through the incorrect entry, dated and initialed by the person making the change. Correction fluid or tape should not be used.

2.4 Electronic records of food safety management activity should be automatically generated and activity logs kept. See Bizmanualz document #ABR34M, procedure ITAD102 – IT Records Management, for guidance.

3.0 FOOD SAFETY RECORD MAINTENANCE

3.1 The authority associated with a particular food safety record (e.g., Production Manager for production records) shall ensure that records are maintained in a suitable environment that prevents damage or deterioration and also prevents loss of records.

- The methods used to prevent loss of hardcopy records may include controlling access, use of checkout cards, or auditing of records as appropriate for the particular record.
- Electronic records are protected from damage, deterioration, loss, or unauthorized change by use of appropriate access control methods. See Bizmanualz document #ABR34M, procedure ITSD106 – IT Access Control, for guidance.

3.2 Active food safety records must be stored such that they are readily retrievable. Active records are those that have not yet met their minimum retention times. Inactive food safety records may be archived. Archived records are retrievable, but not necessarily readily retrievable.

3.3 Unless otherwise indicated, food safety records are destroyed after their minimum retention time is attained, as indicated on FS1010-1, FOOD SAFETY RECORDS LIST.

Effectiveness Criteria:

- Audit results (clear audit trail)

References:

A. Food Safety Manual
   - FSM 4.2.3 – Control of Records

B. Statutory / Regulatory Requirements

The Company should be familiar with – and be able to demonstrate conformance with – the controlling food safety laws in all locations where they participate in the food supply chain. Federal, state, local, and tribal laws may govern the keeping of food safety records by the Company. (An example is the *U.S. Public Health Security and Bioterrorism Preparedness and Response Act of 2002*, or the Bioterrorism Act, section 414 of which pertains to records maintenance and inspection.)

C. Bizmanualz #ABR34M, Computer and Network Policies, Procedures, and Forms (ISBN 1-931591-06-7), 2005.

Records:

- FS1010-1 – FOOD SAFETY RECORDS LIST

Revision History:

Revision	Date	Description of changes	Requested By
0.0	2/1/2006	Initial Release	

[This page intentionally left blank.]

FS1010-1 – FOOD SAFETY RECORDS LIST

Record Title	File Location	Authority or Responsible Party	Minimum Retention	Disposition
Training, education, skills, and experience records ***(incl. Food Safety Team knowledge, experience)***[1]	***Personnel/HR files*** ***Department Files***	***Human Resources Dept.*** ***Department Mgrs.***	***Employment, plus 3 years***	***Destroy***
Quotation Review	Customer Service	Customer Service	1year	Destroy
Sales Order Review	Customer file	Customer Service	3 years	Archive 3 years; Destroy
HACCP Plan Inputs	Project file	Food Safety Team Leader	1 year	Archive 3 yrs.; Destroy
HACCP Plan	***Project file***	***HACCP Coordinator***	***1 year***	***Archive 3 yrs.; destroy***
Supplier Evaluations	Purchasing	Purchasing	5 years	Archive 3 yrs.; Destroy
Purchase Orders	Purchasing	Purchasing	1 year	Archive 3 yrs.; Destroy
Completed Work Orders	Production	Production	3 years	Destroy
Traceability Records	***Food Safety***	***Food Safety***	***1 year***	***Destroy***
Calibration Records	***Quality Assurance***	***Quality Assurance Mgr.***	***2 years***	***Destroy***
Records of nonconforming monitoring/measurement equipment, repair records	***Quality Assurance***	***Quality Assurance Mgr.***	***2 years***	***Destroy***

[1] Records required by the ISO 22000 standard are shown in ***bold italics***.

FS1010-1 – FOOD SAFETY RECORDS LIST

Record Title	File Location	Authority or Responsible Party	Minimum Retention	Disposition
Process monitoring – devices used, monitoring frequency, etc.	*Food Safety*	*Food Safety Team*	*1 year*	*Destroy*
Internal Audit Records	*Administration*	*Management Representative*	*3 years*	*Destroy*
Inspection Records	Food Safety	Food Safety Team	3 years	Archive 7 yrs. Destroy
Product Release Records	Food Safety	Food Safety	1 year	Destroy
Nonconformity Reports	Food Safety	Food Safety Team	3 years	Destroy
Corrective Actions	*Food Safety*	*Food Safety Team Leader*	*5 years*	*Destroy*
Document masters	Document Control	Food Safety	3 years	Destroy
Customer Complaints	Marketing	Marketing	3 years	Destroy
Process validation records	Food Safety	Food Safety Team	2 years	Destroy
Customer satisfaction surveys	Customer Service	Customer Service	3 years	Destroy
Verification of modifications to PRPs	*Food Safety*	*Food Safety Team*	*2 years*	*Destroy*
Input to hazard analyses	*Food Safety*	*Food Safety Team*	*1 year*	*Destroy*
Hazard analysis results	*Food Safety*	*Food Safety Team Leader*	*1 year*	*Archive 1 year, then destroy*

FS1010-1 – FOOD SAFETY RECORDS LIST

Record Title	File Location	Authority or Responsible Party	Minimum Retention	Disposition
Verified process flow diagrams	*Food Safety*	*Food Safety Team*	*2 years*	*Destroy*
Internal/external communication records	*Administration*	*Company Management*	*3 years*	*Destroy*
Records of Management Reviews	*Administration*	*Company Management*	*2 years*	*Archive 3 yrs.; destroy*
Responsibilities & authority of external experts	*Food Safety*	*Food Safety Team*	*2 years*	*Archive 2 yrs.; destroy*
Emergency procedures	*Emergency Mgmt.*	*Emergency Coord./Food Safety Team Leader*	*3 years*	*Destroy*
Emergency logs	*Emergency Mgmt.*	*Emergency Coord./Food Safety Team Leader*	*3 years*	*Archive 2 yrs.; destroy*

[This page intentionally left blank.]

Document #: **FS1020**	Title: **MANAGEMENT RESPONSIBILITY**	Print Date: **2/1/2006**
Rev #: **0.0**	Prepared By:	Date Prepared: **2/1/2006**
Effective Date: **2/1/2006**	Reviewed By:	Date Reviewed: **2/1/2006**
	Approved By:	Date Approved: **2/1/2006**
Standards: **ISO 22000:2005, clause 5**		

Policy: Top Management shall commit to the development, implementation, and continuing improvement of the Company's FSMS.

Purpose: To outline the methods for ensuring that the responsibility, authority and interrelation of all personnel who manage, perform, and verify work affecting food safety is defined and how the Food Safety Management System is to be reviewed to ensure its continuing suitability and effectiveness.

Scope: This procedure applies to the President, Food Safety Team, Food Safety Team Leader, and all department managers involved with food safety.

Responsibilities:

The President holds primary responsibility for implementation of this procedure and, ultimately, the whole Food Safety Management System.

Department Managers, as part of the Management Team, are responsible for supporting the President and the Food Safety Team Leader in the implementation of this procedure.

The Food Safety Team Leader is responsible for managing the Food Safety Team and ensuring that processes needed for the Food Safety Management System are established, implemented and maintained; reporting to the President and Management Team on the performance of the Food Safety Management System and any need for improvement; and ensuring the promotion of awareness of customer requirements pertaining to food safety throughout the organization.

Human Resources (HR) is responsible for establishing and maintaining Job Descriptions to define responsibilities and authorities throughout the company. HR is also responsible for maintaining the Company Organizational Chart.

Definitions: Food Safety Team – Persons responsible for testing, inspecting, and reporting on FSMS procedures, to ensure their conformance to applicable requirements.

Food Safety Team Leader – One who has acquired the necessary competencies, training, certifications, and managerial skills to lead a food safety team.

Management Team – Consists of the Food Safety Team Leader, Department Managers, and the Company President, at a minimum.

Procedure:

1.0 MANAGEMENT PLANNING

1.1 On an annual basis, the President conducts strategic planning meetings with the Management Team before the budgeting process.

1.2 Strategic objectives are set at the strategic planning meetings. Food Safety Objectives, including those needed to meet requirements for products and services, are included. The strategic objectives shall be measurable and the measures defined at the planning session.

1.3 Within one month of the Strategic Planning meeting, Department Managers and the Food Safety Team Leader shall develop departmental objectives in support of the Company's strategic objectives. The departmental objectives shall be specific and measurable, with measurements defined, and shall be consistent with the Company's strategic objectives.

1.4 The President and Management Team shall review the proposed departmental objectives along with the budget requests. The review process may include adjustments to ensure objectives are relevant, challenging, and achievable. The review process shall also ensure resources are in the budget to support the departmental objectives.

1.5 After the departmental objectives are finalized, they are posted on departmental bulletin boards to communicate them to all employees. If any objectives and measures are confidential, they will not be posted or at the least, confidential information will be omitted.

2.0 MANAGEMENT RESPONSIBILITIES AND AUTHORITIES

2.1 The President shall appoint a Food Safety Team Leader, whose responsibilities and authorities (irrespective of any others he/she may have) shall include:

- Managing a Food Safety Team and organizing its work;
- Ensuring relevant training and education of Food Safety Team members;
- Ensuring that the FSMS is established, implemented, maintained, and updated; and
- Reporting to the Company President on the effectiveness, suitability, and continuous improvement of the FSMS.

2.2 The Company President shall appoint a HACCP Coordinator, in accordance with FS1080 – HACCP PLAN MANAGEMENT. Among the HACCP Coordinator's duties will be development and maintenance of the Company's HACCP Plan(s).

2.3 Human Resources, along with Department Managers, shall develop and maintain job descriptions for all Company positions in accordance with FS1040 – JOB DESCRIPTIONS.

2.4 Human Resources shall maintain the Company's organization chart, which should be available to all personnel.

2.5 Management shall appoint a person to serve as a *liaison* between the Company and external parties on matters relating to the Food Safety Management System (see FS1190 – PRODUCT RECALL for an example). At Management's discretion, the Food Safety Team Leader may serve as the Company's external liaison. Experience in the field of public relations may be a desirable qualification (see FS1040 – JOB DESCRIPTIONS).

3.0 MANAGEMENT REVIEW

3.1 At least twice per year (more often, if deemed necessary), The Food Safety Team Leader will coordinate a meeting to be attended by the President and Management Team. The purpose of the meeting is to review the Company's Food Safety Management System, to ensure its continuing suitability, adequacy, and effectiveness. This review shall include assessing opportunities for improvement and the need for changes to the system, including the food safety policy and related objectives.

3.2 The Food Safety Team Leader is responsible for ensuring minutes of the meeting are taken and for maintaining the minutes as a Quality Record, in accordance with FS1010 – FOOD SAFETY RECORDS.

3.3 The meeting agenda will include the following reports from the Food Safety Team Leader and appropriate Department Managers:

- Results of audits;
- Customer feedback;
- Process performance and product safety;
- Status of preventive and corrective actions;
- Follow-up actions from previous management review;
- Changes that could affect the food safety management system; and
- Recommendations for improvement.

3.4 The meeting minutes shall also include any decisions and actions for follow-up at the next review, related to:

- Improvement of the effectiveness of the food safety management system and its processes;
- Improvement of product safety related to customer requirements; and
- Resource needs.

3.5 Management review action items should clearly state who (by name and title) will be responsible for what action(s) and by what date.

Effectiveness Criteria:

N/A

References:

A. Food Safety Manual
- FSM 5.0 – Management Responsibility

B. Food Safety Procedures
- FS1010 – FOOD SAFETY RECORDS
- FS1040 – JOB DESCRIPTIONS

Records:

- Management Review Minutes

Revision History:

Revision	Date	Description of changes	Requested By
0.0	2/1/2006	Initial Release	

Doc #: **FS1030**	Title: **COMPETENCE, AWARENESS, AND TRAINING**	Print Date: **2/1/2006**
Rev #: **0.0**	Prepared By:	Date Prepared: **2/1/2006**
Eff. Date: **2/1/2006**	Reviewed By:	Date Reviewed: **2/1/2006**
	Approved By:	Date Approved: **2/1/2006**
Standards: **ISO 22000:2005, clause 6.2.2**		

Policy: To ensure that all Company employees are competent, trained in the latest food safety principles and techniques, and continually upgrading their knowledge and skills.

Purpose: This procedure outlines the methods and responsibilities for ensuring personnel performing work affecting the safety of our products and services are competent based on education, training, skills, and experience.

Scope: This procedure applies to all Company personnel.

Responsibilities:

Supervisory personnel are responsible for defining the minimum competency requirements for positions under their supervision.

Human Resources is responsible for verifying qualifications indicated on resumes or applications of new hires. HR maintains records supporting competency determination in personnel files.

Procedure:

1.0 NEW EMPLOYEE SELECTION

1.1 The Company shall establish the necessary competencies for personnel whose activities may have an impact on food safety.

1.2 When the need for additional food safety personnel has been established, Human Resources shall screen applications and resumes to select candidates with the appropriate competencies for positions to be filled. See FS1030-1 – FOOD SAFETY TRAINING REQUIREMENTS LIST for guidance.

1.3 Hiring supervisors and managers interview candidates to select personnel that are best qualified for the position.

1.4 Human Resources may verify application or resume information of the final candidates by checking references or previous employers.

1.5 If the best candidate does not have the minimum competence required, the hiring manager shall arrange for training or other actions, as appropriate, to satisfy the requirements for the position.

1.6 The hiring supervisor or manager shall, within a reasonable period following the hiring, evaluate the effectiveness of the training or other action taken. Various methods may be used for evaluation; for example tests, observed performance, or oral interview.

1.7 The supervisor or manager, with Human Resources, shall take appropriate action to modify the training program or provide additional training, if the evaluation shows the training was ineffective.

2.0 NEW EMPLOYEE ORIENTATION

2.1 The department manager is responsible for arranging orientation and initial job training for new employees. HR provides orientation training. Department personnel provide initial job training. At a minimum, orientation covers the following topics:

- Review and understanding of the Company's process maps (flow diagrams);
- Review and understanding of the Company's food safety manual and food safety procedures;
- Review and understanding of the Company's food safety policy and objectives;
- Overview of the Company's products; and
- Review and understanding of workplace rules, guidelines, environmental, and health and safety instructions.

2.2 Initial job training is given to new hires as well as personnel transferring into a new area. The training should cover the following topics, at a minimum:

- Specific training on how to perform the work activities of the job function;
- Review of the relevance and importance of the new employee's activities and how they contribute to the achievement of the food safety objectives; and
- The importance and awareness of effective internal communication, as stated in section 5.6.2 of ISO 22000:2005.

Human Resources shall maintain a record of initial job training for every employee in FS1030-2 – FOOD SAFETY TRAINING LOG.

2.3 To ensure consistent training is provided to all employees in a particular position, defined training guidelines or work instructions shall be used as a basis for the training.

2.4 Department managers are responsible for ensuring competencies are maintained in light of new products and services, new technologies, and changing conditions. When needs are identified, the department manager arranges for in-house training or utilizes HR to identify appropriate outside training.

- The manager evaluates the effectiveness of the training or other action taken.
- Various methods may be used for evaluation; for example tests, observed performance, or verbal interview.

2.5 HR shall maintain a record of all employee training, education, skills and experience in personnel files (see FS1030-2 – FOOD SAFETY TRAINING LOG for guidance). Department managers may keep copies of in-house or job training, but one copy shall be forwarded to HR for updating the master personnel file.

3.0 ONGOING TRAINING

3.1 As changes in the business environment, business requirements, food safety laws, and safety knowledge and techniques dictate, Company employees dealing with food safety shall continue to receive training with regard to food safety techniques and issues.

Company employees whose work directly or indirectly affects food safety shall undergo periodic evaluations of competence and training with regard to the FSMS (this should be done annually, at a minimum).

3.2 Human Resources, in conjunction with the Food Safety Team Leader, shall develop and maintain a training plan for all food safety employees, in accordance with the Company's employee education policy and the FSMS. As stated in 2.5, above, Human Resources shall maintain a record of employee food safety training/education, experience, and skills (see FS1030-2).

3.3 The Food Safety Team Leader, in conjunction with Human Resources, shall periodically evaluate the Company's food safety training & skills requirements with respect to industry standards and best practices, to ensure that employees' food safety skills and knowledge continue to meet business, legal/ regulatory, and customer requirements.

3.4 Human Resources shall periodically review the FS1030-2 – FOOD SAFETY TRAINING LOG to ensure that all employees are receiving the required food safety training on a timely basis and to evaluate the training program, to ensure that the food safety training program is meeting Company and statutory/ regulatory requirements.

References:

A. Food Safety Manual
   - FSM 6.2.2 – Competence, Awareness, and Training

B. Food Safety Procedures
   - FS1040 - Job Descriptions

Records:

- FS1030-1 – FOOD SAFETY TRAINING REQUIREMENTS LIST
- FS1030-2 – FOOD SAFETY TRAINING LOG
- Training Effectiveness Evaluations

- Human Resources / Personnel Files (employee status and relevant education, training, skills, and experience for the position)

Revision History:

Revision	Date	Description of changes	Requested By
0.0	2/1/2006	Initial Release	

FS1030-1 – FOOD SAFETY TRAINING REQUIREMENTS LIST

Employee ID:	
Job Title:	
Grade:	
Skill / Experience Requirements For This Job:	
Next Job Title In Recommended Career Path:	
Next Job Grade:	
Skill/ Experience Requirements For Next Job:	

Recommended Career Path:

(example:

- packer
- grader
- grading supervisor
- quality analyst
- quality supervisor
- etc.)

[This page intentionally left blank.]

FS1030-2 – FOOD SAFETY TRAINING LOG

Employee Name: ____________________

Employee ID: ____________ **Department:** ____________

Employee Title / Grade: ____________________

COURSE ID	PREREQ COURSE ID	CERTIF. COURSE (Y/N)	START DATE	END DATE	PASS / NO PASS

HR Management: ____________________ **Date:** ____________

Food Safety Team Leader: ____________________ **Date:** ____________

[This page intentionally left blank.]

Document #: **FS1040**	Title: **JOB DESCRIPTIONS**	Print Date: **2/1/2006**
Revision #: **0.0**	Prepared By: **preparer's name, title**	Date Prepared: **2/1/2006**
Effective Date: **2/1/2006**	Reviewed By: **reviewer's name, title**	Date Reviewed: **2/1/2006**
	Approved By: **name, title**	Date Approved: **2/1/2006**
Standards: **ISO 22000:2005, clause 6.2.2**		

Policy: The Company shall identify necessary competencies for employees whose activities impact food safety.

Job descriptions, including the necessary food safety competencies, will be prepared for all Company positions, to serve as an organizational aid for identifying and delegating responsibilities, coordinating and dividing work, and preventing duplication of efforts.

Purpose: To provide the methods for preparation and format of Job Descriptions. A Job Description should be used as a guide. Job descriptions are not intended to be all-inclusive of a person's abilities, the requirements for fulfilling their position, or as work limitations or restrictions on employee roles. We are all expected to be team players and to help each other and the Company whenever necessary, within reason and workplace safety guidelines.

Scope: Applies to all managers and the Human Resources department.

Responsibilities:

Human Resources is responsible for assuring that current job descriptions, meeting all pertinent legal requirements, are developed and maintained for all positions at Our Company.

Supervisory Personnel are responsible for defining food safety competencies and creating job descriptions for all positions within their respective scopes of responsibility.

The HR Manager acts as the ADA Coordinator and is responsible for assisting hiring managers in composing job descriptions which comply with Americans with Disabilities Act (ADA) requirements regarding the creation of job descriptions.

Definition: Disability – Under the ADA, a disability is defined as a "physical or mental impairment that substantially limits one or more major life activities, such as seeing, hearing, speaking or working, etc." Illiteracy is even covered if caused by a physical or mental disorder.

Procedure:

1.0 JOB DESCRIPTION PREPARATION

1.1 Hiring managers should be responsible for initiating drafts or changes to existing job descriptions. Whenever practical, supervisors should interact with other employees – managerial and non-managerial – when developing or reviewing descriptions in order to better ensure accurate, clear, and meaningful job descriptions.

1.2 Job descriptions should be current and appropriately represent the position and the needs of the Company at all times.

- Job descriptions should be updated whenever positions, reassignment of duties, organizational changes, etc., are required. Job descriptions should mirror the growth and changes of the company. Hiring managers should not fall into a routine of allowing individuals or Company operations to be governed by pre-existing descriptions.

1.3 The HR Manager should work in conjunction with hiring managers to finalize job descriptions.

2.0 JOB DESCRIPTION FORMAT AND CONTENT

Job descriptions should be prepared using FS1040-1 – JOB DESCRIPTION as a guide. Any unusual needs or requirements for the position should be added in a separate section. The job description format is explained in the following paragraphs.

2.1 Job Title

The title represents the name of the position. It should be short and simple, yet as descriptive as possible.

2.2 Effective Date

This is the latest revision date to the description or the implementation date.

2.3 Department

The department name should be clear and accurately identify the department function. Alternatively, it could identify the Department Code along with the Department Name.

2.4 Summary of Functions

This should be a one or two sentence statement encompassing the basic function and objectives of this position, so the applicant or employee can grasp at a glance the key reasons why the job exists. It should enable anyone reasonably familiar with the organization to understand the primary purpose of the position. Any constraints, particular emphasis, or shared responsibility can also be mentioned.

2.5 Major Duties and Responsibilities

This section should briefly describe specific job tasks with details of the major duties and/or responsibilities for performing the job. Whenever possible use descriptive terms related to the objectives or action of a particular function rather than to indicate merely what is done. It is very important to note specific deliverables for a task.

It is recommended that job descriptions contain no more than ten duties. Ideally, five to six duties will be included in order to make the responsibilities description easier to understand. Try to describe all the specific aspects of each job in a short space. Related tasks or activities should be grouped together to describe what is to be achieved. Usually, higher level jobs use broader duty statements.

2.6 Food Safety Responsibilities

How the position affects safety of the Company's products – *and* how the position is affected by food safety issues – must be clearly and concisely addressed in the job description. If the person in the given position is responsible for one or more critical control points, this must be addressed within the HACCP plan (see FS1080 – HACCP PLAN MANAGEMENT) and the Job Description must state all applicable HACCP plan requirements.

2.7 Organizational Relationships

This section should outline the reporting relationships between this position and other key positions including supervisors and positions supervised. This statement should also include the requirements for coordination with other positions or departments.

2.8 Qualifications

If applicable, indicate the minimum requirements necessary to be able to fill the position. For example, this can include a description of the minimum years of experience or accomplishments in specific job categories; degrees/diplomas from colleges, technical, or trade schools; and/or certifications required to perform the job satisfactorily.

2.9 Physical Demands

Use this section to list the typical demands for applicants. Explain the primary demands that require physical and/or mental activities with enough detail in order to make a reasonable distinction for workers compensation issues that may arise.

The Physical Demands section can assist the company in identifying the most qualified applicant available for a specific job based on reasons unrelated to a disability. The documented demands of a job can be used as a basis for Americans with Disabilities Act (ADA) compliance or workers' compensation claims.

2.10 Work Environment

Describe the environment of the position. Explain any special circumstances involving the physical area that may be important (e.g., Is the work environment noisy or quiet? Indoors or out? Office or factory? Unheated or not air conditioned? Physically challenging (see 2.8)?)

3.0 JOB DESCRIPTION APPROVAL AND DISTRIBUTION

3.1 The next higher management level approves Job Descriptions prepared by their subordinate hiring managers. For example, the Board of Directors should approve the President's job description and those of his direct reports.

3.2 Human Resources shall be responsible for communicating completed job descriptions to hiring managers and their superiors. Job descriptions should be included with the respective department's organization chart in the Organization Structure section of the Company's policy-and-procedure manual.

3.3 Human Resources shall maintain the signed, official hardcopy of all job descriptions in HR files and ensure the availability of job descriptions to all Company employees (e.g., via the Company intranet).

4.0 JOB DESCRIPTION REVIEW

4.1 As part of an annual management review, Human Resources shall review job descriptions to aid Top Management in determining if jobs and job descriptions accurately reflect the Company's short- and long-term requirements, as well as meet statutory/regulatory, food safety, and customer requirements.

4.2 Supervisory personnel create or develop revisions to job descriptions in accordance with the first three sections of this procedure and submit such descriptions to Human Resources for review.

4.3 Human Resources, in cooperation with Top Management, shall indicate necessary changes to job descriptions and return to the responsible supervisor.

4.4 The responsible supervisor shall incorporate needed changes and resubmit the descriptions to Top Management for final approval. HR shall then file the revised/new descriptions for posting, as those positions become available.

Effectiveness Criteria:

- No obsolete job descriptions on file with HR.
- All personnel understand what is expected of them.

Additional Resources:

A. None.

References:

A. Food Safety Manual
    - FSM 6.2.2 – Competence, Awareness, and Training

B. Food Safety Procedures

- FS1030 – COMPETENCE, AWARENESS, AND TRAINING

C. Statutory/Regulatory Requirements

The U.S. Americans with Disabilities Act of 1990 gives civil rights protections to individuals with disabilities, similar to those protections provided to individuals on the basis of race, color, sex, national origin, age, and religion. It guarantees equal opportunity for individuals with disabilities in public accommodations, employment, transportation, state and local government services, and telecommunications.

The Title I employment provisions apply to private employers with 15 or more employees, state and local governments, employment agencies, and labor unions. The ADA prohibits discrimination in all employment practices, including job application procedures, hiring, firing, advancement, compensation, training, and other terms, conditions, and privileges of employment. It applies to recruitment, advertising, tenure, layoff, leave, fringe benefits, and all other employment-related activities.

The Company is required to be aware of – and operate within – the statutes and regulations governing jobs and hiring in all places where it conducts business.

Records:

- FS1040-1 – JOB DESCRIPTION

Revision History:

Revision	Date	Description of changes	Requested By
0.0	2/1/2006	Initial Release	

[This page intentionally left blank.]

FS1040-1 – JOB DESCRIPTION

Title Effective Date

Department

Summary Of Functions

Major Duties And Responsibilities

Food Safety Requirements/Responsibilities

Organizational Relationships

Qualifications

Physical Demands

Work Environment

Rev. # _________ Effective Date __________________

Approved (Dept. Mgr.) ________________________________ Date: _____________

Approved (HR) ________________________________ Date: _____________

Approved (Top Mgmt.) ________________________________ Date: _____________

[This page intentionally left blank.]

Doc #: **FS1050**	Title: **PREREQUISITE PROGRAMS**	Print Date: **2/1/2006**
Revision: **0.0**	Prepared By:	Date Prepared: **2/1/2006**
Eff. Date: **2/1/2006**	Reviewed By:	Date Reviewed: **2/1/2006**
	Approved By:	Date Approved: **2/1/2006**
Standard(s): **ISO 22000:2005, clauses 7.2.1, 7.2.2, 7.2.3, and 7.5**		

Policy: The Company will develop, implement, and maintain prerequisite programs (PRPs) to control food safety hazards within its scope of operations.

Purpose: To prevent or eliminate food safety hazards or to minimize the likelihood of food safety hazards being introduced through the work environment; to control biological, chemical, and physical contamination of the product(s), including cross-contamination between products; and to control food safety hazard levels in the product and product processing environment.

Scope: This procedure applies to any Company function or department that may have an impact – directly or indirectly – on food safety.

Responsibilities:

The Food Safety Team is responsible for developing the Company's prerequisite program(s) and for ensuring that PRPs are implemented and monitored.

The Food Safety Team Leader (FSTL) is responsible for putting together a Food Safety Team (or teams), ensuring PRPs are communicated to the appropriate parties, and periodically reviewing PRPs.

Department employees are responsible for carrying out their duties in accordance with the relevant PRP(s) and reporting on their PRP-related work.

Department Managers are responsible for assigning PRP-related duties, ensuring that employees have adequate training to carry out their PRP work, and reviewing PRP logs to identify deviations (nonconformities) and potential hazards.

Definitions: Cross-contamination – The transfer of harmful bacteria from one food or food ingredient to another by way of a nonfood surface, such as a cutting board, countertop, utensils, or a preparer's hands.

Food safety hazard – Biological, chemical, or physical agent in food, or condition of food, with the potential to cause an adverse health effect.

Hazard analysis – The process of collecting and evaluating information on hazards associated with the food under consideration to decide which are significant and must be addressed in the HACCP plan.

Material Safety Data Sheet (MSDS) – Provides workers and emergency personnel with the proper procedure(s) for handling or working with a particular substance. MSDS's include information such as physical data (melting point, boiling point, flash point, etc.), toxicity, health effects, first aid, reactivity, storage, disposal, protective equipment, and spill/leak procedures that are of particular use if a spill or other accident occurs.

Operational prerequisite program (operational PRP) – A PRP identified during a hazard analysis as essential to controlling (a) the likelihood of introducing food safety hazards to, (b) contamination of, or (c) proliferation of food safety hazards in the product(s) or in the processing environment.

Prerequisite program (PRP) – Basic conditions and activities necessary to maintain a hygienic environment throughout the food supply chain which is suitable for production, handling, and provision of safe end products and safe food for human consumption. Examples of prerequisite programs include various "good practices" within the food supply chain (e.g., Good Agricultural Practice, Good Veterinary Practice, Good Storage Practice, Good Inventory Practice, Good Manufacturing Practice, Good Sanitation Practice, etc.).

Procedure:

1.0 PREREQUISITE PROGRAMS – BACKGROUND

1.1 Prerequisite programs – PRPs – are the basic conditions and activities needed to maintain a hygienic environment throughout the food supply chain. Every organization in the food supply chain is required to have its own set of PRPs that address food safety issues in its unique environment

1.2 The Company shall have its own set of PRPs. Which PRPs the Company implements will depend on the nature of its business and its place in the food safety chain. These PRPs should be based on established "good practices". Examples of good practices include, but are not limited to:

- Good Agricultural Practice;
- Good Distribution Practice;
- Good Handling Practice;
- Good Hygienic Practice;
- Good Maintenance Practice;
- Good Manufacturing Practice;
- Good Pest Control Practice;

- Good Production Practice;
- Good Sanitation Practice;
- Good Storage Practice;
- Good Trading Practice; and
- Good Veterinary Practice.

1.3 PRPs are established *in advance of* any hazard analysis (see FS1060 – HAZARD ANALYSIS PREPARATION and FS1070 – HAZARD ANALYSIS). One *result* of a hazard analysis is the establishment of certain PRPs as *operational* PRPs (see section 5.0).

2.0 PREREQUISITE PROGRAM PLANNING

2.1 The Food Safety Team Leader (FSTL) shall establish a team of individuals knowledgeable in the department/activity and specialists in the department/ activity in question, at a minimum. This group shall be known as the Food Safety Team for that department / activity.

- The FSTL should consider consulting regulatory authorities and quality assurance specialists, as needed, to expand the knowledge base of the Food Safety Team.

2.2 The Food Safety Team shall develop prerequisite programs using FS1050-1 – PREREQUISITE PROGRAM EXAMPLE as a guide.

2.3 The Food Safety Team Leader shall ensure the departmental/activity PRP is communicated accurately and thoroughly to employees responsible for implementing it in a timely manner.

3.0 IMPLEMENTING PREREQUISITE PROGRAMS

3.1 Depending primarily on the nature of the Company's business, it may operate more than one PRP (see Additional Resource "A" for guidance).

3.2 Each PRP shall consist of:

- One or more procedures, designed to ensure safety in the processing, handling, etc., of food and/or food ingredients (see FS1050-2 – STANDARD OPERATING PROCEDURE (SOP) FORM - EXAMPLE);
- Training requirements for the procedure(s);
- Forms, lists, records, etc., required to carry out the procedure(s).

See FS1050-1 – PREREQUISITE PROGRAM EXAMPLE and FS1050-5 – PRP LOG EXAMPLE for additional guidance.

3.3 The Department Manager shall assign primary responsibilities for carrying out a given PRP. Persons to whom PRP responsibilities have been assigned shall maintain PRP work logs and report to the Department Manager at least once a day on all PRP-related work, to ensure that deviations (nonconformities) are identified and promptly acted on, in accordance with FS1140 – CONTROL OF MONITORING AND MEASURING and FS1150 – CONTROL OF NONCONFORMING PRODUCT.

3.4 The Department Manager shall ensure that employees are adequately trained to handle their PRP duties and responsibilities in accordance with FS1030 – COMPETENCE, AWARENESS, AND TRAINING.

4.0 REVIEWING PREREQUISITE PROGRAMS

4.1 The Food Safety Team Leader (or an appointed member of the Food Safety Team) shall periodically review PRPs to determine if they continue to meet Company, regulatory, and customer requirements and continue to produce the desired results. Such reviews should be conducted at least semiannually.

4.2 PRPs should be subjected to third-party audits at regular intervals – even if such audits are not legally required – to verify that the PRPs are properly documented and communicated to the appropriate parties, properly implemented, continue to meet requirements, and are monitored and measured for the purpose of continual improvement.

5.0 ESTABLISHING OPERATIONAL PREREQUISITE PROGRAMS

5.1 Operational prerequisite programs are PRPs that have been identified during a hazard analysis (see FS1070 – HAZARD ANALYSIS) as ***essential*** to controlling:

- The likelihood of introducing food safety hazards to;
- Contamination of; or
- Proliferation of food safety hazards in

the Company's product(s) or processing environment.

5.2 PRPs identified as *operational* PRPs in the hazard analysis process shall be documented. Such documentation shall include the following, at a minimum, for each operational PRP:

- Food safety hazards to be controlled;
- Control measures;
- Monitoring procedures to demonstrate that operational PRPs have been implemented;
- If monitoring shows that an operational PRP is not in control, corrections and corrective actions to be taken in accordance with FS1150 – CONTROL OF NONCONFORMING PRODUCT;
- Responsibilities and authorities; and
- Monitoring records.

Effectiveness Criteria:

- PRP logs (see FS1050-5) are filled out completely at the end of every working day (or shift) and are up-to-date.
- Standard operating procedures (see FS1050-2) are followed and tasks are performed according to schedule.
- Timely review and revision (if needed) of PRPs.

Additional Resources:

A. Hazard Analysis And Critical Control Point Principles And Application Guidelines, National Advisory Committee On Microbiological Criteria For Foods, Food & Drug Administration, U.S. Department of Agriculture, 1997.

B. Developing Food Safety Systems Manual for Retail Meat Operations, Beef Information Centre, Canadian Meat Council, 2003.

References:

A. Food Safety Manual

- FSM 7.2 – Prerequisite Programs
- FSM 7.2.4 – Establishing Operational Prerequisite Programs

B. Statutory / Regulatory Requirements

While various food industry standards (such as ISO 22000) require prerequisite programs, regulatory requirements for PRPs may vary according to locality. The Company should be aware of and observe applicable laws governing food safety and address all pertinent legal requirements within their PRPs.

Certain ***aspects*** of PRPs are regulated in every country and locality, to some extent. For example, the U.S. Food Code specifies elements of PRP's – such as cooking and storage temperatures – for many products. The Company is advised to maintain awareness of food-safety-related legislation pertaining to its line(s) of business. This can be done by seeking qualified legal advice. The Company can also maintain ongoing contact with – and actively participate in – trade associations, industry councils, extension services, and advisory boards.

C. Food Safety Procedures

- FS1030 – COMPETENCE, AWARENESS, AND TRAINING
- FS1060 – HAZARD ANALYSIS PREPARATION
- FS1070 – HAZARD ANALYSIS
- FS1140 – CONTROL OF MONITORING AND MEASURING
- FS1150 – CONTROL OF NONCONFORMING PRODUCT

Records:

- FS1050-1 – PREREQUISITE PROGRAM EXAMPLE

- FS1050-2 – STANDARD OPERATING PROCEDURE (SOP) FORM - EXAMPLE
- FS1050-3 – EXAMPLE APPROVED CHEMICALS / AUTHORIZED HANDLERS LIST
- FS1050-4 – STORAGE MAP EXAMPLE
- FS1050-5 – PRP LOG EXAMPLE

Revision History:

Revision	Date	Description of changes	Requested By
0.0	2/1/2006	Initial Release	

FS1050-1 – PREREQUISITE PROGRAM EXAMPLE

(NOTE: While this example and the following forms are based on a "sanitation PRP", the Company must have a PRP in place for every one of its operations that has an effect on the safety of its end products.)

1. **SANITATION SOPs**
   Cleaning and sanitation procedures are described on Standard Operating Procedure forms (see FS1050-2 – EXAMPLE SOP FORM), which will include a list of chemicals used, sanitation procedures and frequencies, and identify the person(s) responsible for carrying out such procedures.

2. **STORING CHEMICALS**
   - All cleaning / sanitation chemicals are stored in well-ventilated areas and in their original labeled containers.
     If chemicals are temporarily placed in other than their original containers (i.e., dispensed from a gallon jug into a half-liter spray bottle for ease of use), the other container must be clearly labeled to prevent mixing of incompatible chemicals.
   - Chemical storage areas, indicated on the Storage Map (see FS1050-4 – STORAGE MAP EXAMPLE), are adequately separated from food preparation and food storage areas.

3. **SELECTING AND HANDLING CHEMICALS**
   - All chemicals for cleaning and sanitation are suitable for use in food establishments and are approved by the appropriate food safety regulatory agency (i.e., US FDA).
   - Material Safety Data Sheets (MSDS) are kept on-site for all chemicals used in cleaning and sanitation activities.
   - All chemicals used for cleaning / sanitation are noted on an Approved Chemicals and Authorized Chemical Handlers list (see FS1050-3 – APPROVED CHEMICALS / AUTHORIZED HANDLERS LIST EXAMPLE). Persons responsible for applying or mixing chemicals are trained by qualified personnel and are identified on the same list.

4. **CLEANING / SANITATION EQUIPMENT**
   - Non-disposable cloths, when used in the sanitation program, are cleaned and disinfected before every use.
   - Brushes are inspected before each use to ensure bristles are not loose.
   - Hoses are fitted with valves (nozzles) to prevent water from entering. When not in use, hoses are evacuated and stored off the floor, rolled up to prevent stagnant water from accumulating inside.

5. **ROOM TEMPERATURE**
   If the air temperature in a production area exceeds 50°F (10°C) for more than four consecutive hours, a mid-shift cleanup of the area is performed.

6. PROTECTING FOOD (INGREDIENTS) WHILE CLEANING / SANITIZING

- During cleaning and sanitizing, all food, food ingredients, and packaging materials are covered and/or relocated to prevent chemicals from contacting them.
- During cleaning operations, care is taken to avoid water splashing from the floor onto clean surfaces. Cleaned surfaces are kept free of excess pooled water to prevent growth of microorganisms.
- Sanitizers and other chemicals are rinsed from all surfaces, unless specifically indicated as a "no-rinse" treatment.

7. TRAINING

- All individuals performing cleaning / sanitation activities are trained by qualified personnel.
- All individuals performing cleaning / sanitation activities are required to read and submit a signed copy of the Sanitation Program and the Sanitation Standard Operating Procedures forms for the areas for which they are responsible (a) at the start of employment and (b) following any changes to procedures and policies.

8. RECORD OF CLEANING / SANITATION ACTIVITY AND CORRECTIVE ACTIONS

- Activities related to the cleaning/sanitation program, as well as any corrections or corrective actions required, are recorded in the (FS1050-5) SANITATION LOG by the person designated by the Department Manager as they are performed.
- The Department Manager reviews the Sanitation Log each day and reports deviations immediately to the Food Safety Team Leader. The Department Manager submits a weekly report on the Sanitation Log to the Food Safety Team Leader.

FS1050-2 – STANDARD OPERATING PROCEDURE (SOP) FORM - EXAMPLE

Our Company – Meat Department	
Item / Description	
Object/Area: Cutting board	**Location:** Meat cutting room
Pre-operational Sanitation	
Responsible party:	Fred Freeman, sanitation crew
Name(s) of chemical(s), cleaning product(s) used and conditions of (specifications for) use:	• Brand "X" cleaner – dilute 2.0 centiliters in 1.0 liter of water • Brand "Y" sanitizer – dilute 1.0 milliliter in 1.0 liter of water
Procedure:	• Remove cutting board to cleaning area. • Hose down board with warm water. • Apply cleaner to one side of board. Allow to stand for ten (10) minutes and rinse. Repeat for other side. • Apply sanitizer to one side of board. Allow to stand for two (2) minutes and rinse. Repeat for other side. • Return board to meat cutting room.
Frequency/timing:	Every day, prior to the start of a shift.
Operational Sanitation	
Responsible party:	Meat cutting personnel
Name(s) of chemical(s), cleaning product(s) used and conditions of (specifications for) use:	None
Procedure:	Turn over cutting board halfway through an 8-hour shift.
Frequency/timing:	Every day, at the halfway point of an 8-hour shift (at 12:00 pm for an 8:00 am – 4:00 pm shift)

SOP Creation Date: ____________________ Approved By (Mgmt.): ____________________

Last Update: ____________________ Updated By: ____________________

[This page intentionally left blank.]

FS1050-3 – APPROVED CHEMICALS / AUTHORIZED HANDLERS LIST EXAMPLE

Our Company – Meat Department				
Chemical Name	***Chemical Manufacturer***	***FDA Approval Code***	***Authorized Handler(s)***	***Approved Use***
Odorono	StaneFree, Inc.	ABC123	Willy Loman	surface sanitizing

Creation Date: mm/dd/yyyy

Last Update: mm/dd/yyyy

Approved By (Mgmt.): ______________________

Updated By: ______________________

[This page intentionally left blank.]

FS1050-4 – STORAGE MAP EXAMPLE

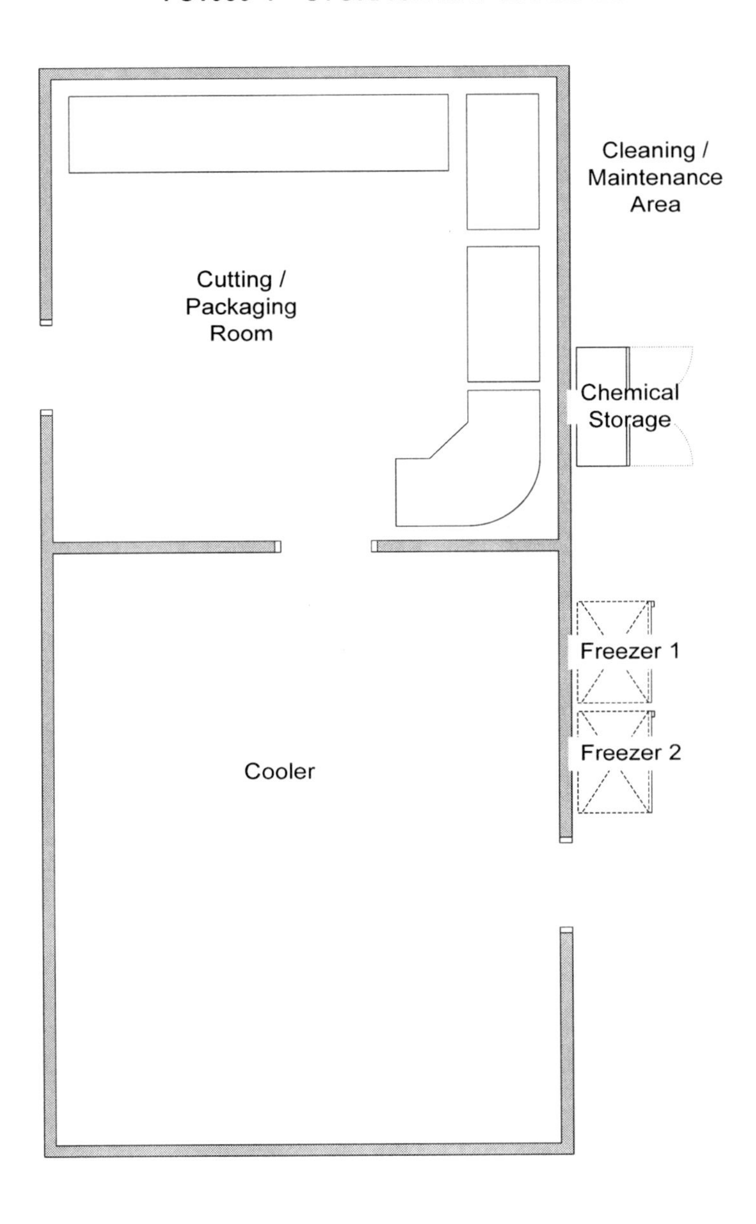

[This page intentionally left blank.]

FS1050-5 – PRP LOG EXAMPLE

Our Company – Meat Department **SANITATION**		
Chemical Use	***Yes***	***No***
Chemicals used are suitable for food establishments and are shown on the Approved Chemicals / Authorized Handlers List	☐	☐
Persons using cleaning / sanitizing chemicals are appropriately trained and are identified on the Approved Chemicals / Authorized Handlers List	☐	☐
Chemical Storage		
Cleaning / sanitizing chemicals are stored away from food products and food contact surfaces and stored where indicated on the Storage Map (See FS1050-4)	☐	☐
Cleaning / sanitizing chemicals are clearly labeled	☐	☐
Cleaning / sanitizing chemicals are properly sealed when not in use and containers are in good condition (are not leaking, corroding, or deteriorating)	☐	☐
Cleaning / sanitizing equipment		
Water pressure and temperature are adequate	☐	☐
Concentrations of cleaning / sanitizing chemicals are adequate	☐	☐
Necessary cleaning / sanitizing equipment is readily available in adequate numbers (amounts)	☐	☐
Equipment is in good working order and using it will not present a hazard	☐	☐
Cleaning / sanitation SOP's		
Sanitation SOP forms were followed for all equipment and facilities, which are ready for production	☐	☐
Wherever necessary, food products and food packaging materials were covered, protected, and/or removed from the area being cleaned / sanitized	☐	☐
Any equipment disassembled for cleaning / sanitizing has been reassembled, inspected, and found to be in good working order	☐	☐

For every "No" above, describe the corrective action(s) taken
Corrective action completed – verified by: ____________________ ***Date:*** ______________ *(mm/dd/yyyy)*

Form Completed On: mm/dd/yyyy

Completed By: ____________________

Form Verified On: mm/dd/yyyy

Verified By: ____________________

[This page intentionally left blank.]

Doc #: **FS1060**	Title: **HAZARD ANALYSIS PREPARATION**	Print Date: **2/1/2006**
Rev #: **0.0**	Prepared By:	Date Prepared: **2/1/2006**
Eff. Date: **2/1/2006**	Reviewed By:	Date Reviewed: **2/1/2006**
	Approved By:	Date Approved: **2/1/2006**
Applicable Standard(s): **ISO 22000:2005, clause 7.3**		

Policy: The Company shall collect, maintain, update, and document all information needed to conduct hazard analyses.

Purpose: By ensuring accurate, thorough, and effective hazard analyses, the Company is better able to guarantee the safety and efficacy of food products and/or ingredients under its control and ensure that its products conform to customer and statutory/regulatory requirements.

Scope: This procedure applies to all Company personnel involved in the hazard analysis process.

Responsibilities:

The Food Safety Team is responsible for preparing a Flow Diagram for each product or process category covered by the Food Safety Management System and for verifying the accuracy of those diagrams.

The Food Safety Team Leader is responsible for selecting members of the Food Safety Team and managing them throughout the hazard analysis planning process.

Human Resources is responsible for maintaining records of training and qualifications for members of the Food Safety Team.

The Legal Department (or qualified legal representative of the Company) is responsible for maintaining an awareness of statutory/regulatory requirements pertaining to food safety and for advising the Company of those requirements.

Top Management is responsible for appointing a Food Safety Team Leader to conduct/manage hazard analyses.

Definitions: Food Safety Management System (FSMS) – An ordered, well-documented system, designed to ensure consistency and improvement of work procedures and practices, including produced goods, and result in safe food.

Food Safety Team – Personnel responsible for testing, inspecting, and reporting on Food Safety Management System (FSMS) procedures to ensure their conformance to applicable requirements.

Hazard Analysis – Process of collecting and evaluating information on hazards associated with a food ingredient/product, to determine which hazards are significant and must be addressed in the HACCP plan. Hazard analysis consists of two steps, identification and evaluation.

Safe food – Food that is not harmful when consumed; food that does not cause medical illness or pose a health hazard to the consumer.

Procedure:

1.0 THE FOOD SAFETY TEAM

1.1 Depending on the size and nature of the Company's business, Top Management may either:

- Appoint a Food Safety Team and Food Safety Team Leader to conduct each individual hazard analysis; or
- Establish a full-time, managerial level position of Food Safety Team Leader (or Food Safety Manager), who shall select and manage a Food Safety Team for the duration of a hazard analysis.

1.2 The Food Safety Team shall have a combination of multidisciplinary knowledge and experience in developing and implementing the Food Safety Management System. Such knowledge/experience shall include – but may not be limited to – the Company's products, processes, equipment, and food safety hazards that fall within the scope of the Food Safety Management System.

1.3 Human Resources shall maintain rccords that demonstrate the Food Safety Team possesses the required knowledge and experience, in accordance with FS1030 – COMPETENCE, AWARENESS, AND TRAINING.

2.0 PRODUCT CHARACTERISTICS

2.1 The Food Safety Team shall document *all raw materials, ingredients, and product contact materials* (e.g., processing equipment, conveyors, packaging) to the extent necessary to conduct a hazard analysis (see FS1070 – HAZARD ANALYSIS). Such documentation shall include, but not limited to, the following:

- Biological, chemical, and physical characteristics of raw materials/ingredients;
- Composition of formulated ingredients (e.g., additives and processing aids);
- Point of origin of all materials;
- Method of production;
- Packaging and delivery methods;
- Storage conditions and shelf life;
- Preparation and/or handling *before* use or processing; and

- Food-safety-related acceptance criteria or specifications of purchased materials and ingredients appropriate to their intended uses.

2.2 The Food Safety Team shall document characteristics of *end products* to the extent necessary to conduct a hazard analysis (see FS1070 – HAZARD ANALYSIS). Such documentation shall include, but not be limited to, information on the following:

- The product name – brand name and/or common product description (e.g., chili beans, tenderloin, whole wheat bread) – and related information (e.g., SKU);
- Composition (ingredients and amounts of each);
- Biological, chemical, and physical characteristics relevant to food safety;
- Intended shelf life and storage conditions;
- Packaging, especially that which comes into direct contact with the product;
- Labeling related to food safety and/or instructions for handling, preparation, and usage; and/or
- Method of distribution.

2.3 The Legal Department (or Company-authorized legal advisor) shall identify statutory and regulatory requirements related to product characteristics. Descriptions shall be kept up-to-date, including, when required, modifications done in accordance with FS1080 – HACCP PLAN MANAGEMENT.

3.0 INTENDED USE

3.1 The Food Safety Team shall consider and document all intended uses, reasonably expected handling, and any unintended – but reasonably expected – *mis*handling and *mis*use of end products to the extent needed to conduct a hazard analysis (see FS1070 – HAZARD ANALYSIS).

3.2 The Food Safety Team shall identify and document groups of users/consumers of each product. Consumer groups known to be especially vulnerable to specific food safety hazards (e.g., allergies to foods or food ingredients, like peanuts or dairy products) shall be considered.

3.3 Intended-use documentation shall be kept up-to-date, including modifications done in accordance with FS1080 – HACCP PLAN MANAGEMENT, when required.

4.0 FLOW DIAGRAMS, PROCESS STEPS, AND CONTROL MEASURES

4.1 The Food Safety Team shall prepare an FS1060-1 – FLOW DIAGRAM EXAMPLE for each product or process category covered by the FSMS. Flow diagrams provide a basis for evaluating any possible occurrence, increase, or introduction of food safety hazards.

4.2 Flow diagrams shall be clear, accurate, and sufficiently detailed and shall include the following, *as appropriate*:

- The sequence and interaction of all steps in the operation;
- Any outsourced processes and subcontracted work;
- Where raw materials, ingredients, and intermediate products enter the flow;
- Where rework and recycling take place; and
- Where end products, intermediate products, by-products, and waste are released or removed.

5.0 VERIFICATION

5.1 The Food Safety Team shall verify the accuracy of flow diagrams and other hazard analysis documents by conducting process walkthroughs and on-site checks.

5.2 Verified flow diagrams shall be maintained in accordance with FS1010 – FOOD SAFETY RECORDS.

Effectiveness Criteria:

- Organized, efficient hazard analyses
- Hazard analyses performed quickly and accurately

References:

A. Food Safety Manual

- FSM 7.3 – Hazard Analysis Preparation

B. Statutory / Regulatory Requirements

There are laws in virtually every country and in most localities requiring producers to sell food that is safe to consume. Many governmental bodies have legislation in place that prescribes hazard analysis and HACCP plans.

The Company is required to know and observe all applicable laws pertaining to safety of their end products.

C. Food Safety Procedures

- FS1010 – FOOD SAFETY RECORDS
- FS1030 – COMPETENCE, AWARENESS, AND TRAINING
- FS1070 – HAZARD ANALYSIS
- FS1080 – HACCP PLAN MANAGEMENT

Records:

- FS1060-1 – FLOW DIAGRAM EXAMPLE

Revision History:

Revision	Date	Description of changes	Requested By
0.0	2/1/2006	Initial Release	

[This page intentionally left blank.]

FS1060-1 – FLOW DIAGRAM EXAMPLE

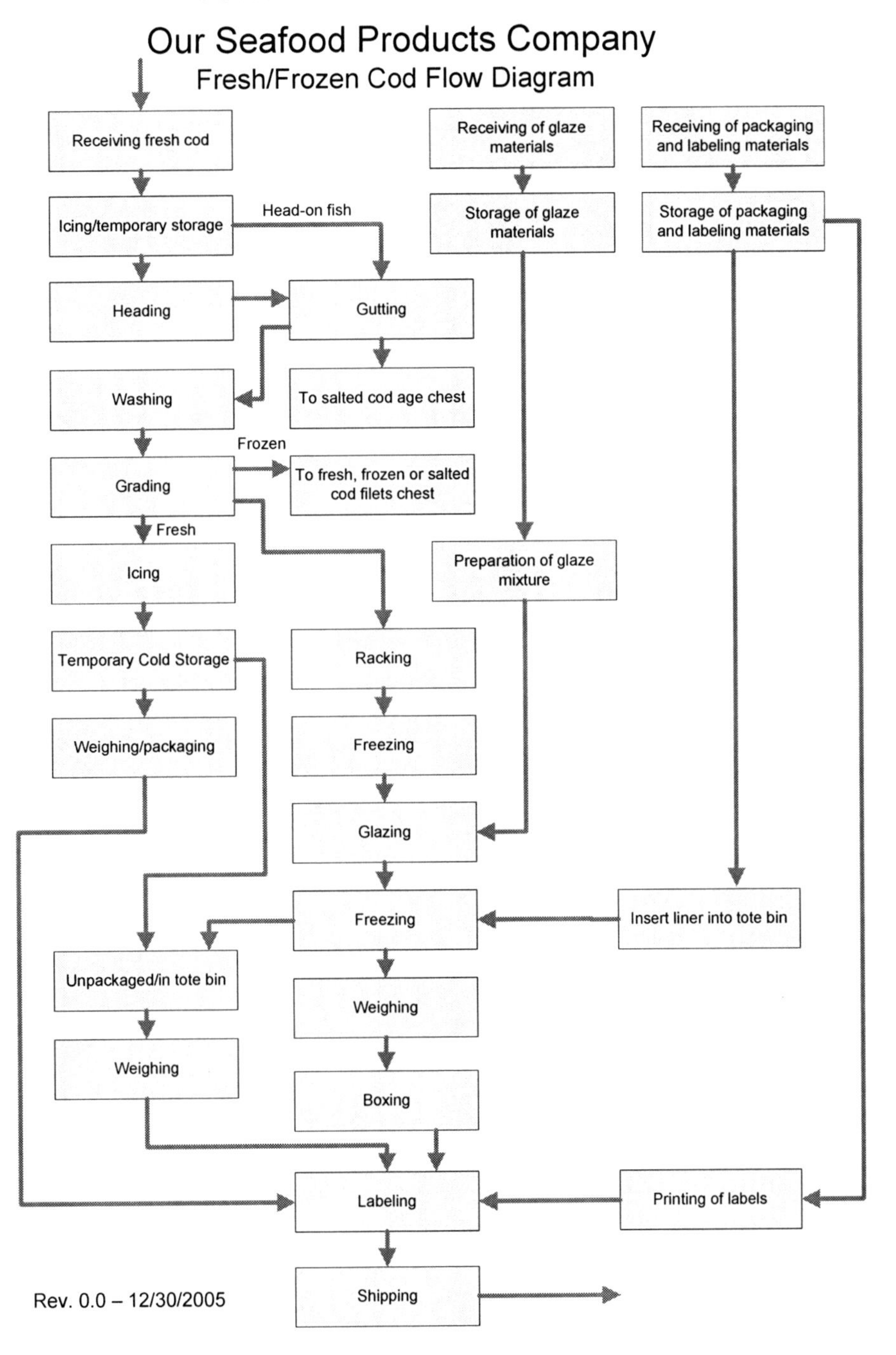

[This page intentionally left blank]

Doc #: **FS1070**	Title: **HAZARD ANALYSIS**	Print Date: **2/1/2006**
Rev #: **0.0**	Prepared By:	Date Prepared: **2/1/2006**
Effective Date: **2/1/2006**	Reviewed By:	Date Reviewed: **2/1/2006**
	Approved By:	Date Approved: **2/1/2006**
Standards: **ISO 22000:2005, clause 7.4**		

Policy: The Company shall conduct hazard analyses to identify critical control points in its food handling and processing operations.

Purpose: To outline the steps necessary for conducting effective hazard analyses that conform to statutory / regulatory requirements and the Company's HACCP Plan and, most importantly, result in safe end products.

Scope: This procedure applies to all Company food products and to any process related to food safety.

Responsibilities:

The Food Safety Team is responsible for conducting hazard analyses.

The Food Safety Team Leader is responsible for supervising hazard analyses, reporting results of such analyses to the rest of the Management Team, and periodically reviewing food-safety-related processes to ensure hazards are accounted for and controlled.

The Management Team (consisting of Managers of affected departments and chaired by the Food Safety Team Leader) is responsible for ensuring implementation of effective hazard controls, assisting in development of food safety process revisions, as necessary, and for implementing revisions to those processes.

The Quality Assurance Manager is responsible for helping to improve hazard controls and food-safety-related processes.

Definitions: Acceptable level – The level of a safety hazard considered to present, at worst, a risk the consumer would accept. The acceptable level of the final product, sometimes referred to as the "target level", should be stated in the product description and set at or below any regulatory limits.

Critical Control Point (CCP) – A step at which control can be applied and which is essential to prevention or elimination of food safety hazards or reduction of that hazard to an acceptable level. Examples include cooking, mixing, transporting, and storing.

Control measure – An action performed in order to prevent a food safety hazard, eliminate it, or reduce it to an acceptable level.

Flow Diagram – A schematic, systematic representation of the sequence and interactions of steps in a process. A flow diagram usually takes the form of a flowchart, where all steps in a process and their inputs and outputs (including byproducts and waste) are shown as boxes connected by (mostly) unidirectional arrows. Flow diagrams are sometimes referred to as "process maps".

Food Safety Hazard – Biological, chemical, or physical agent in food, or condition of food, with the potential to cause an adverse health effect. Not to be confused with the term "risk", which, in the context of food safety, is a function of the *probability* of an adverse health effect (e.g., becoming diseased) and the *severity* of that effect (death, hospitalization, absence from work, etc.) when exposed to a specified hazard.

Hazard Analysis – The process of collecting and evaluating information on hazards associated with the food under consideration to decide which are significant and must be addressed in the HACCP plan. A hazard analysis consists of two steps, identification and evaluation.

High-risk food – A food that supports the growth of bacteria and/or microbes, such as meat, dairy, or eggs.

Monitoring – Regular measurement or observation of a critical control point to make sure a product does not go outside of its critical limits.

Prerequisite program (PRP) – Universal step or procedure that controls operational conditions within a food establishment, allowing for environmental conditions favorable to the production of safe food. An example of a PRP is cleaning food processing equipment after every use.

Target – Standard which must be met in order to control a hazard.

Procedure:

1.0 HAZARD ANALYSIS – GENERAL INFORMATION

1.1 The Food Safety Team shall conduct a hazard analysis of each of the Company's processes that potentially have an impact on food safety, to determine:

- Which hazards – biological (e.g., *Salmonella*), chemical (e.g., mercury, dioxin), and/or physical (e.g., bones, metal) – need to be controlled;
- The degree of control required to ensure food safety; and
- Which control measures are required.

1.2 In conducting a hazard analysis, wherever possible the Food Safety Team should include the following:

- The likelihood of hazards and severity of their adverse health effects;
- The qualitative and/or quantitative evaluation of the presence of hazards;
- Survival or multiplication of microorganisms of concern;

- Production or persistence in foods of toxins, chemicals, or physical agents; and
- Conditions leading to any of the above.

1.3 The Food Safety Team shall review all relevant process maps / flow diagrams (see FS1060-1 – FLOW DIAGRAM EXAMPLE) in order to become familiar with the Company's processes in advance of a hazard analysis.

2.0 HAZARD IDENTIFICATION AND DETERMINATION OF ACCEPTABLE LEVELS

2.1 The Food Safety Team shall use relevant tools and knowledge as means of identifying food safety hazards that are reasonably expected to occur in relation to the type of product, type of process, and processing facilities. Means of hazard identification may include, but not be limited to:

- Preliminary information and data collected in accordance with FS1060 – HAZARD ANALYSIS PREPARATION;
- Available knowledge and experience of the FST members;
- External information, such as epidemiological, historical, regulatory, and other data;
- Information from within the food chain on safety hazards that are or may be relevant to the safety of the Company's end products and the food at the end of the food chain; and
- Flow diagrams (process maps) indicating the step(s) – from raw materials and ingredients to distribution to customers – at which each food safety hazard may be introduced (see 1.3).

2.2 When identifying the hazards, the Food Safety Team shall consider:

- The steps preceding and following the specified operation;
- The process equipment, utilities/services, and surroundings; and
- The preceding and following links in the food supply chain.

2.3 The FST shall record all known and identified hazards in the *Hazards* column of FS1070-1 – HAZARD ANALYSIS CHECKLIST.

2.4 For each of the food safety hazards identified, the Food Safety Team shall determine an *acceptable level* of the food safety hazard in the end product, whenever possible. The determined level shall take into account established statutory and regulatory requirements, customer food safety requirements, the intended use by the customer, and other relevant data. The Food Safety Team shall record justification for – and the result of – the determination in the *Controls and Targets* column of the FS1070-1.

3.0 HAZARD ASSESSMENT

3.1 For each food safety hazard identified on the FS1070-1 – HAZARD ANALYSIS CHECKLIST, the Food Safety Team shall conduct a hazard assessment to determine whether:

- Its elimination or reduction to acceptable levels is essential to the production of a safe food; and
- Its control is needed to enable the defined acceptable levels to be met.

This information shall be recorded in the *Monitoring* column of the FS1070-1 form.

3.2 Each food safety hazard shall be evaluated according to the possible severity of adverse health effects and the likelihood of their occurrence. The methodology used shall be described and the results of the food safety hazard assessment shall be recorded in the *Assessment* column of the FS1070-1.

4.0 SELECTION AND ASSESSMENT OF CONTROL MEASURES

4.1 The Food Safety Team shall identify an appropriate combination of control measures capable of preventing, eliminating, or reducing these food safety hazards to defined acceptable levels. Each control measure (described in FS1060 – HAZARD ANALYSIS PREPARATION) shall be reviewed with respect to its effectiveness against the identified food safety hazards and recorded in the *Control Measures* column of FS1070-1.

4.2 The FST shall determine whether the control measure needs to be managed through operational PRPs or by the HACCP plan, using a logical approach that may include the following kind of tool:

	PRP	OPRP	HACCP
Effect on identified food safety hazards relative to the strictness applied.	No	No	Yes
Its feasibility for monitoring (e.g., ability to be monitored in a timely manner that enables immediate corrective actions).	No	Yes	Yes
Its place within the system relative to other control measures.	No	No	Yes
The likelihood of failure in the functioning of a control measure or significant processing variability.	No	No	Yes
The severity of the consequence(s) in the case of failure in its functioning.	No	No	Yes
Whether the control measure is specifically established and applied to eliminate or significantly reduce the level of hazard(s).	No	Yes	Yes
Synergistic effects (i.e., interaction that occurs between two or more measures resulting in their combined effect being higher than the sum of their individual effects).	No	No	Yes

4.3 Control measures identified as belonging to a HACCP Plan shall be implemented in accordance with FS1080 – HACCP PLAN MANAGEMENT. Control measures not belonging to a HACCP Plan shall be implemented as operational PRPs, in accordance with FS1050 – PREREQUISITE PROGRAMS.

4.4 The Food Safety Team shall document the methodology and parameters used for the categorization and record the results in the *PRP/HACCP* column of FS1070-1.

5.0 HAZARD ANALYSIS REVIEW

5.1 The Food Safety Team Leader shall periodically (at least once every six months – more frequently if a hazard has greater potential for severity or if the likelihood of it occurring is greater than normal) review all processes potentially affecting the safety of the Company's food products, to ensure that all hazards are accounted for and are being controlled.

5.2 If a process needs improving to ensure food safety hazards are minimized or eliminated *or* if regulations or standards affecting the process have been established or revised, the Food Safety Team Leader shall work with the Quality Assurance Manager and the affected Department Manager(s) to improve the process. The revised process shall be tested prior to implementation.

5.3 Within ten (10) days of implementation of the revised process, the Food Safety Team Leader shall review activity during the last ten days to ensure that the process has been properly communicated and implemented *and* to ensure that the process, as revised, is yielding the expected results.

Effectiveness Criteria:

- Potential food hazards are identified and controlled.

Additional Resources:

A. National Agricultural Law Center, University of Arkansas – Fayetteville, Fayetteville, AR, USA (http://www.nationalaglawcenter.org/readingrooms/foodsafety/).

B. Stirling Council Environmental Health – Hazard Analysis Guide (www.stirling.gov.uk/hazard_analysis_guide.doc).

C. South Holland District Council Hazard Analysis Checklist & Advisory (http://www.sholland.gov.uk/website/pdf/eh/food/FAwardChecklistinfo.pdf).

References:

A. Food Safety Manual

- FSM 7.4 – Hazard Analysis

B. Statutory / Regulatory Requirements

In nearly every country, food producers are required by law to develop and follow a Hazard Analysis and Critical Control Point (HACCP) plan for every process potentially affecting the safety of their food products. The Company must recognize and observe the federal, state, and local laws that govern each location in which it does business. An example of law governing hazard analysis is Regulation (EC) number 852/2004 of the European Parliament of the Hygiene of Foodstuffs, which requires companies doing business in the EU to put in place food safety management procedures based on HACCP principles.

C. Procedure Sections

- FS1060 – Hazard Analysis Preparation
- FS1080 – HACCP Plan Management

Records:

- FS1070-1 – HAZARD ANALYSIS CHECKLIST (EXAMPLE)

Revision History:

Revision	Date	Description of changes	Requested By
0.0	2/1/2006	Initial Release	

FS1070-1 – HAZARD ANALYSIS CHECKLIST

(Note: The following is an example of a hazard analysis is that might be conducted on a deli-lunch counter. The Company must have a checklist for each of its processes that affect food safety in any way, in accordance with the Company's HACCP Plan.)

Time and Temperature Control

a) Identify the areas (processes) where temperature control is essential for food safety.

b) Put in place a control and monitoring system to ensure that the required temperature is always met.

A. ACTIVITY (What is taking place?)	B. HAZARDS (What can go wrong?)	C. CONTROLS AND TARGETS (What can be done about the hazard?)	D. MONITORING (How can we check?)	E. ASSESS-MENT	F. CONTROL MEASURES	G. PRP/HACCP
Receiving food product	*Salmonella* growth on chilled, ready-to-eat food if held above required temperature	Person receiving goods to check temperature immediately on delivery: • If above 40°F (5°C), reject. • If at or below 40°F, place in appropriate storage refrigerator immediately, to await processing or further inspection.	Enter temperature on receiving log (chart), initial, and note date/time. Supervisor to review chart weekly and initial.	(ex., "Food temp. 38°F on receipt. Moved to storage pending addl. inspection.")	• Portable digital thermometer • Periodic thermometer calibration	PRP #123 – Receiving Chilled RTE Foods

(Hazard Analysis Checklist begins on next page.)

FS1070-1 – HAZARD ANALYSIS CHECKLIST

A. ACTIVITY	B. HAZARDS	C. CONTROLS AND TARGETS	D. MONITORING	E. ASSESS-MENT	F. CONTROL MEASURES	G. PRP/HACCP
Receiving goods						
Storing high-risk foods (meats, cheeses, eggs, etc.)						
Food on display (e.g., sandwiches, meats at counter)						
Defrosting (poultry, etc.)						
Preparation (e.g., time kept outside of refrigerator)						
Cooking (roasting, frying, microwaving, etc. – prep times, temperatures by weight)						

A. ACTIVITY	B. HAZARDS	C. CONTROLS AND TARGETS	D. MONITORING	E. ASSESS-MENT	F. CONTROL MEASURES	G. PRP/HACCP
Cooling						
Hot holding						
Reheating						
Delivering goods to customers						
Using raw eggs; making products containing raw egg material (e.g., mayonnaise)						
Storage						

[This page intentionally left blank.]

Doc #: **FS1080**	Title: **HACCP PLAN MANAGEMENT**	Print Date: **2/1/2006**
Rev #: **0.0**	Prepared By:	Date Prepared: **2/1/2006**
Effective Date: **2/1/2006**	Reviewed By:	Date Reviewed: **2/1/2006**
	Approved By:	Date Approved: **2/1/2006**
Standards: **ISO 22000:2005, clauses 7.6, 7.7, and 7.8**		

Policy: The Company shall develop, implement, and maintain a HACCP Plan for each of its food operations to help ensure the safety of its products.

Purpose: To delineate the steps needed to develop, implement, and maintain each of the Company's HACCP plans and to help the Company develop HACCP plans that meet statutory, regulatory, and customer requirements.

Scope: This procedure pertains to all processes that directly or indirectly affect the safety of the Company's food products.

Responsibilities:

The HACCP Coordinator is responsible for leading development, implementation, and maintenance of the Company's HACCP Plan(s).

The HACCP Team is responsible for developing, ensuring implementation of, and maintaining the Company's HACCP system. (At a minimum, the HACCP Team is comprised of the Food Safety Team Leader and the QA Manager.)

All employees involved in processes affecting food safety are responsible for implementing the HACCP Plan(s).

Top Management is responsible for committing to the HACCP Plan(s), communicating its ongoing commitment to the Plan(s), and for giving final approval to the HACCP Plan(s).

Definitions: Food safety hazard – Biological, chemical, or physical agent in food - or condition of food - with the potential to cause an adverse health effect. Food safety hazards include allergens (e.g., peanuts).

HACCP – ***H***azard ***A***nalysis and ***C***ritical ***C***ontrol ***P***oint – a systematic approach to identification, evaluation, and control of food safety hazards.

HACCP Plan – Written document, based on HACCP principles, that delineates food safety procedures to be followed by the Company.

HACCP System – Result of implementing the Company's HACCP Plan.

HACCP Team – Persons responsible for developing, implementing, and maintaining the Company's HACCP system.

Procedure:

1.0 DEVELOPING A HACCP PLAN

1.1 Top Management shall appoint a HACCP Team, consisting of persons having knowledge and expertise to develop a HACCP plan and including diverse disciplines (including but not limited to production, inspection, sanitation, microbiology, and engineering). From the HACCP Team members, one person shall be selected as the HACCP Coordinator, who shall supervise the HACCP Team's efforts.

- The HACCP Team should include personnel directly involved with day-to-day processing activities, for their familiarity with variability and limitations of the process.
- The HACCP Team may require some outside expertise (e.g., a public health expert) but should not consist solely of outside personnel, nor should the Company's HACCP Plan(s) be developed entirely outside the Company.

1.2 The HACCP Team shall develop a complete description of the Company's product(s), to assist in identifying possible hazards inherent in the product, the ingredients, or the materials in which they are packaged, in accordance with FS1060 – HAZARD ANALYSIS PREPARATION

- The Company's product description should include, but may not be limited to:

  a. Generic product name (e.g., "egg yolk mix");
  b. Important end product characteristics (e.g., pH, preservatives, color, texture, odor);
  c. Intended use of the product (e.g., "ingredient used in producing baked goods");
  d. The type of packaging, including materials and packaging conditions (e.g., "5-gallon round plastic container, lined with 2-mil plastic bag");
  e. Shelf life and prescribed storage conditions (e.g., "one year if stored at or below 30°F (-1°C)");
  f. Where sold (e.g., "food processors");
  g. Labeling instructions (e.g., "keep frozen"); and
  h. Special distribution controls (e.g., "transport in freezer trailer").

  The HACCP Team shall identify the product's intended use (c and f, above) based on its *normal* use by the customer. They shall indicate where the product will be sold and to whom or what (the target), noting if the target is a special case / segment of the population (e.g., elder persons, infants, groups with dietary prescriptions).

- The Company's product ingredient and incoming material description should include, but may not be limited to:

  a. All raw materials and ingredients used in the making of the product (e.g., whole eggs, preservatives, flavoring);

b. Packaging materials (e.g., styrofoam, plastic wrap, corrugated cardboard); and
c. Processing aids (e.g., stabilizer).

1.3 The HACCP Team shall develop and verify a flow diagram for each of the Company's product processes, in accordance with FS1060 – HAZARD ANALYSIS PREPARATION.

1.4 The HACCP Team shall develop a HACCP Plan for each product process, based on the seven HACCP principles:

- PRINCIPLE 1 – List food safety hazards and possible control measures associated with each step in the flow diagram(s) in columns 2 and 3 of FS1080-1 – HACCP PLAN WORKSHEET, in accordance with FS1070 – HAZARD ANALYSIS.

- PRINCIPLE 2 – For each hazard identified in the hazard analysis, list ***critical control points*** – measures in place to prevent, eliminate, or reduce the hazard to an acceptable level (e.g., cooking, chilling, formulation control) – in column 1 of FS1080-1 – HACCP PLAN WORKSHEET. NOTE: A ***decision tree*** (see next page) may help you determine critical control points.

- PRINCIPLE 3 – Establish ***critical limits*** for each CCP identified and record in column 4 of FS1080-1 – HACCP PLAN WORKSHEET. A "critical limit" is what separates acceptable products from unacceptable ones.

  a. Determine critical limits for each CCP.

  Such information is often established and readily available (from scientific publications, regulatory guidelines, experimental studies, or food safety experts, for example). In some cases, the appropriate critical limit may not be readily available and testing (by the Food Safety Team Leader, QA Manager, or an outside expert) may be required.

  An example – when cooking beef patties, critical limits may include minimum internal temperature of the patties, the oven temperature, time in the oven, patty thickness, and moisture content.

  b. Validate all critical limits under actual conditions.

- PRINCIPLE 4 – Establish monitoring procedures for each CCP, to determine if the operation is within critical limits at that point.

  a. Determine ***what will be monitored*** and record this information in column 5 of FS1080-1 – HACCP PLAN WORKSHEET. This is usually a measurement or observation (e.g., measuring storage temperature or processing line speed, or verifying that shellfish were harvested from approved waters).

Codex Decision Tree – Principle 2

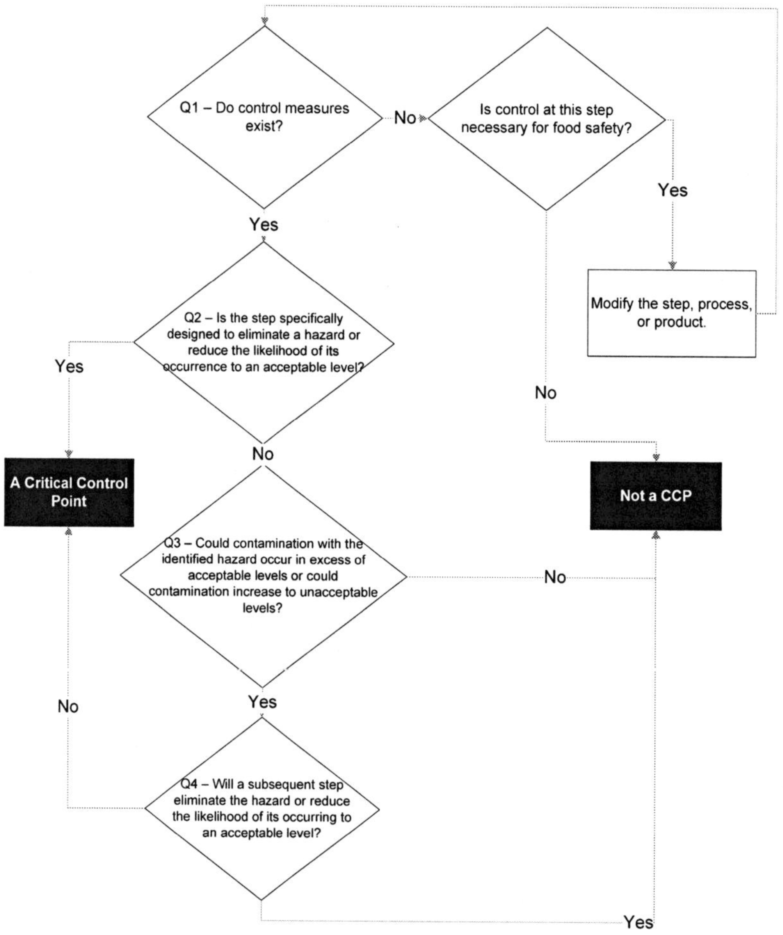

Q1 – Do control measures exist?

No

Is control at this step necessary for food safety?

Yes

Modify the step, process, or product.

No

Yes

Q2 – Is the step specifically designed to eliminate a hazard or reduce the likelihood of its occurrence to an acceptable level?

Yes

No

A Critical Control Point

Not a CCP

Q3 – Could contamination with the identified hazard occur in excess of acceptable levels or could contamination increase to unacceptable levels?

No

Yes

No

Q4 – Will a subsequent step eliminate the hazard or reduce the likelihood of its occurring to an acceptable level?

Yes

b. Determine ***how critical limits and preventive measures will be monitored and recorded*** and record this information in <u>column 6</u> of FS1080-1. Physical or chemical properties (e.g., temperature, pH, time) are measured or observations recorded and instant, real-time results must be provided: lengthy testing and analysis are inappropriate in food safety, as the product could be distributed before results are received.

All monitoring equipment must be accurate and precise enough to measure critical limits correctly. Periodic standardization or calibration of instruments, in accordance with FS1140 – CONTROL OF MONITORING AND MEASURING, is required.

c. Determine and record ***monitoring frequency*** and record this information in column 7 of FS1080-1. Frequency of monitoring must be appropriate to the CCP and critical limits being monitored. Monitoring may be ***continuous*** or ***intermittent***. Examples of continuous monitoring include recording time and temperature on a temperature recording chart or computerized log and using a metal detector.

   If monitoring is continuous, such equipment used must be checked periodically to make sure it is operating correctly (see FS1140 – CONTROL OF MONITORING AND MEASURING).

   If continuous monitoring is not possible, intermittent measurements (e.g., sampling batter temperature on a breading line) must be taken at appropriate intervals. Intermittent monitoring frequency will depend on such factors as the degree of process variance, evidence of variability trends, proximity of measured values to critical limits, and the level of risk the Company is willing to assume.

d. Determine ***who will be responsible*** for monitoring each CCP and record this information in column 8 of FS1080-1 – HACCP PLAN WORKSHEET. Line personnel, equipment operators, and/or Quality Assurance personnel may be responsible for monitoring a given process. Every employee responsible for monitoring a CCP must:

   - Be trained to perform the specific monitoring activity at that CCP (see FS1030 – COMPETENCE, AWARENESS, AND TRAINING);
   - Fully understand the importance of CCP monitoring;
   - Have full and ready access to the monitoring activity;
   - Accurately report each monitoring activity; and
   - Immediately report any deviation / nonconformity, so that appropriate corrective action (see FS1170 – CORRECTIVE ACTION) may be taken.

- PRINCIPLE 5 – Establish planned corrections and corrective actions (deviation handling) for each CCP and record these in column 10 of FS1080-1.

  A process operator shall take corrective action, in accordance with FS1170 – CORRECTIVE ACTION, whenever monitoring indicates a process is operating outside of specified critical limits.

  When the process is found operating outside of critical limits, the operator will apply a correction by handling the potentially unsafe product in accordance with FS1150 – CONTROL OF NONCONFORMING PRODUCT.

- PRINCIPLE 6 – Establish verification steps for each monitoring activity and record this information in column 11 of FS1080-1.

Each monitoring activity must include additional methods or tests to verify the HACCP system is working as intended.

Verification shall be performed according to a preset schedule and less frequently than the monitoring activity being verified. Verification may include:

a. A periodic review of records generated while monitoring;

b. Calibration of monitoring equipment; and

c. Supplemental testing (e.g., microbiological assays).

- PRINCIPLE 7 – Establish required records and recordkeeping/documentation procedures and record them in columns 9 and 10 of FS1080-1 – HACCP PLAN WORKSHEET.

  Generating and maintaining records is ***mandatory*** for all CCP monitoring and verification activity. Necessary records must be identified, which must include all pertinent information and be signed and dated by the person in charge of monitoring and reviewing the process (see column 8, FS1080-1), in accordance with FS1010 – FOOD SAFETY RECORDS.

2.0 IMPLEMENTING A HACCP PLAN

2.1 The HACCP Coordinator shall produce the HACCP Plan in a format like the one shown in FS1080-2 – HACCP PLAN OUTLINE and submit the Plan to Top Management for approval.

2.2 Once the HACCP Plan is approved, the HACCP Coordinator shall introduce the Plan to all Company employees, to build awareness of the Plan and promote understanding of the Plan's importance to the Company, its customers, and the consumer at the end of the supply chain.

2.3 The HACCP Coordinator shall ensure that employees involved in any process having an effect on food safety (e.g., receiving, processing, packaging, cleaning) receive appropriate HACCP Plan training prior to the Plan's implementation, in accordance with FS1030 – COMPETENCE, AWARENESS, AND TRAINING.

2.4 Prior to implementation, the HACCP Coordinator shall conduct a "mock implementation" to verify the various aspects of the Plan and allow ample time to make changes to the Plan in advance of the implementation deadline.

2.5 Once the HACCP Plan has been tested and verified, it can be implemented.

3.0 HACCP PLAN REVIEW

3.1 The HACCP Coordinator shall review the Company's HACCP Plan(s) on a regular basis – annually, at a minimum – to determine if it continues to meet Company, customer, and statutory/regulatory requirements and continues to result in a safe end product.

3.2 The HACCP Plan(s) shall be periodically subjected to a third-party audit (the period may be determined by statute – if not, a third-party audit should be conducted at least annually), to verify that the Plan is clear, sound, and continues to meet Company, customer, and statutory/regulatory requirements AND that the Plan results in a safe product.

3.3 In the event of a food safety emergency (e.g., unsafe product found at final inspection, product recalled), the employee discovering the emergency shall initiate a corrective action in accordance with FS1170 – CORRECTIVE ^CTION. The HACCP Coordinator should periodically review corrective actions to determine if the HACCP Plan requires revision.

4.0 HACCP PLAN REVISION

4.1 After a HACCP Plan review or a review of corrective actions, the HACCP Coordinator shall ensure updating of the Plan, when and where needed.

4.2 Once updated, the Plan shall be tested and verified and the test results, with the revised HACCP Plan, shall be submitted to Top Management for review and approval.

4.3 Within a week of updating a HACCP Plan, Top Management shall ensure that the update has been tested and verified (i.e., necessary tests have been conducted and documented and the results inspected (by a qualified outside inspector, if required)). The HACCP Coordinator shall retain such inspection records in accordance with FS1000 – DOCUMENT CONTROL and FS1010 – FOOD SAFETY RECORDS.

Effectiveness Criteria:

- Improved food safety processes.
- Fewer deviations from prescribed critical limits.
- Fewer product recalls.
- Consistent, assured conformance to statutory / regulatory and customer requirements.
- Continual improvement of the Company's HACCP Plan(s).

Additional Resources:

A. FDA/CFSAN Food Code 2005, Public Health Service – US Food & Drug Administration, Department of Health & Human Services, 2005 – http://www.cfsan.fda.gov/~dms/foodcode.html#get05

B. Food Safety Enhancement Program Implementation Manual, Volume 2 – Guidelines and Principles for Development of HACCP Generic Models (2nd Ed.), Canadian Food Inspection Agency (CFIA), October, 2000.

C. Hazard Analysis Critical Control Point (HACCP) Information Center, Iowa State University, Ames, IA, USA – http://www.iowahaccp.iastate.edu/.

D. Ensuring Safe Food – A HACCP-Based Plan for Ensuring Food Safety in Retail Establishments, Extension Service Bulletin 901, The Ohio State University, Columbus, OH, USA – http://ohioline.osu.edu/b901/index.html.

E. Guidance for Industry – Juice HACCP Hazards and Controls Guidance (1st ed., 2004), US FDA/CFSAN – http://www.cfsan.fda.gov/~dms/juicgu10.html

References:

A. FSM Manual

- FSM 7.5 – Managing the HACCP Plan
- FSM 7.6 – Updating the Food Safety Management System

B. Statutory / Regulatory Requirements

Since the HACCP concept was developed in the 1960's, it has been a valuable tool, a Good Practice, and a strategic differentiator for many companies. It has had such a positive impact that many governments have made or are considering making HACCP plans a legal requirement for food companies. Examples of existing HACCP plan regulations include:

- U.S. Code of Federal Regulations, Title 9, Chapter 3, Part 417 (or "9 CFR 417") – Hazard Analysis and Critical Control Point (HACCP) Systems; and
- Fish Inspection Regulations, Consolidated Regulations of Canada (CRC), c. 802.

C. Food Safety Management Procedures

- FS1000 – DOCUMENT CONTROL
- FS1010 – FOOD SAFETY RECORDS
- FS1030 – COMPETENCE, AWARENESS, AND TRAINING
- FS1060 – HAZARD ANALYSIS PLANNING
- FS1070 – HAZARD ANALYSIS
- FS1140 – CONTROL OF MONITORING AND MEASURING
- FS1150 – CONTROL OF NONCONFORMING PRODUCT
- FS1170 – CORRECTIVE ACTION

Records:

- FS1080-1 – HACCP PLAN WORKSHEET
- FS1080-2 – HACCP PLAN OUTLINE

Revision History:

Revision	Date	Description of changes	Requested By
0.0	2/1/2006	Initial Release	

[This page intentionally left blank.]

FS1080-1 – HACCP PLAN WORKSHEET

Cooked Shrimp Operation Example[1]

1. Critical Control Point (CCP)	2. Significant Hazard	3. Control/ Preventive Measure	4. Critical Limits	Monitoring				9. Records	10. Corrections, Corrective Actions, and Records	11. Verification
				5. What?	6. How?	7. How Often?	8. Who?			
Cooking	Survival of Listeria can result in illness or death	Heat Process – 5D[2] Listeria Reduction	2 min. @ 100°C (212°F) will result in internal product temp. of 80°C (176°F) for 1 second	Conveyor belt speed Temperature of cooker Recorder chart	Conveyor speed with stopwatch Recording thermometer Visual inspection	After each break Continuous Hourly	QA staff Automatic QA staff	Conveyor belt monitoring record Recorder Chart QA initials recorder chart	1. Isolate affected product and evaluate for safety. 2. Record in non-conformity Corrective Action log book. 3. Sign and date the Corrective Action taken 4. Determine source of problem, take action to prevent recurrence 5. Retrain employees, if necessary	1. Have the QA Manager verify corrective actions daily. 2. QA to review cook log 3. QA Manager observe cooking process, compare data with data obtained by cooker operator 4.Verify heat process 5. Calibrate temperature recorder

[1] "Fish, Seafood and Production – How to Reengineer Your QMP Plan – A Manual for Fish Processors", Canadian Food Inspection Agency, 2002.
[2] 5D (5 log, or 100,000:1) reduction in *Listeria*.

FS1080-1 – HACCP PLAN WORKSHEET

Company Operation

1. Critical Control Point (CCP)	2. Significant Hazard	3. Control/ Preventive Measure	4. Critical Limits	Monitoring				9. Records	10. Corrections, Corrective Action and Records	11. Verification
				5. What?	6. How?	7. How Often?	8. Who?			

FS1080-2 – HACCP PLAN OUTLINE

1. Cover Page
    a. Company (plant) name and address
    b. Purpose Statement — Detail the purpose of the manual (e.g., "This manual lists the products manufactured at this location and provides a full description of the hazards, preventive measures, corrective actions, and verifications used to produce safe food products.").
    c. Commitment Statement — Expressing Top Management's commitment to initiate and perform the program detailed in this manual (e.g., "By signing this document, the owners of Our Company agree to accept and perform the duties described in the HACCP manual and to empower Company employees to carry out the procedures described in the manual.").
    d. Signatures of Company officials (Top Management) committing to the program.
    e. Date on which the Company will implement the HACCP Plan.
2. HACCP Team Members
    a. HACCP Coordinator's name and contact information
    b. Coordinator's credentials (qualifications, certifications, etc.)
    c. HACCP Team members' names, titles, and contact information
3. List of products covered in the HACCP Plan
4. For each process category or HACCP Plan:
    a. Fully describe the product and method of distribution
    b. Identify the consumer and the intended use of the food product
    c. Provide a flow diagram for each process

    (Items d-k come from each FS1080-1 – HACCP PLAN WORKSHEET)

    d. List the identified hazard(s)
        - State the significance of each hazard
        - Justify (explain) the significance
    e. Describe preventive measures
    f. Identify critical control points (CCP)
    g. Identify critical limits
    h. Describe monitoring procedures and frequencies
    i. Describe corrections and corrective actions to be taken
    j. Describe the record keeping system

   k. Describe verification procedures
5. Employee training records
   a. New hires
   b. Ongoing training (on the job, classes, seminars, certifications)
6. Recall procedure

Doc #: **FS1090**	Title: **PURCHASING**	Print Date: **2/1/2006**
Rev #: **0.0**	Prepared By:	Date Prepared: **2/1/2006**
Eff. Date: **2/1/2006**	Reviewed By:	Date Reviewed: **2/1/2006**
	Approved By:	Date Approved: **2/1/2006**
Standards: **ISO 22000:2005, clause 7.1**		

Policy: The Company shall purchase safe food ingredients and/or products and shall use such products only as intended.

Purpose: To define the methods to be used for procuring materials, supplies, and services at our Company.

Scope: This procedure applies to the purchase of all inventory items, supplies, materials and ingredients, subcontracted services, and capital equipment affecting food safety within the Company.

Responsibilities:

All personnel that require a product or service must complete purchase requisitions specifying items for purchase and obtain required approvals.

Purchasing is responsible for evaluating suppliers, maintaining raw material inventories, placing orders with approved suppliers, negotiating pricing with suppliers, and forwarding all paperwork to accounting for payment.

Accounting and Accounts Payable are responsible for payment of invoices only after satisfactory completion or delivery of goods or services has been made.

Food Safety Management is responsible for including inspection requirements on Purchase Orders.

Receiving and Warehouse Personnel are responsible for receiving, inspecting materials, and forwarding all paperwork to Purchasing.

Procedure:

1.0 ORDER DETERMINATION AND REQUISITION

1.1 Purchasing shall determine reorder quantities of standard production inventory items (e.g., product ingredients) by comparing available stock on hand with the requirements to satisfy the production plan.

1.2 For non-inventory production items, including supplies and engineering components, and services, the originating individual or department shall prepare a FS1090-1 – PURCHASE REQUISITION. Requisitions should be completed and approved with the following items and any additional supporting documentation, as appropriate:

- Complete description with part or model numbers, if available;
- Specifications, including food safety considerations;
- Type, class, and/or grade required;
- Quantity required;
- Date required;
- Requesting department and accounting code;
- Recommended vendor or source, if applicable;
- Food Safety Management System requirements;
- Special shipping requirements; and
- Special inspection requirements upon receipt.

1.3 If the requisition is for subcontracted ***services***, the FS1090-1 must include:

- A complete description of the service to be performed;
- Specifications, if appropriate;
- Requirements for qualification of personnel;
- Food safety standards to be applied; and
- Food Safety Management System requirements.

1.4 Purchasing shall analyze terms, vendor, pricing, quantity breaks, etc., and shall order accordingly in the Company's best interest. Purchasing shall obtain approval from the requestor of any material variances prior to placement of the order.

1.5 Vendor selection for inventory and non-inventory items and subcontracted services shall be conducted in accordance with FS1100 – SUPPLIER EVALUATION. Only suppliers on the FS1100-1 – APPROVED VENDOR LIST may be used for items or services that may affect the safety of Our Company's food products or services.

2.0 ORDER PLACEMENT

2.1 Purchasing shall complete a FS1090-2 – PURCHASE ORDER for all orders. The Purchase Order should be completed with all applicable information (see purchase requisition data above), authorized by the appropriate parties, and entered into the FS1090-3 – PURCHASE ORDER LOG.

Purchase Orders for standard inventory items print directly from the computer system using data from the Item Master. Inspection requirements identified by Food Safety Management will show up as comments on the Purchase Order.

- Purchasing should include arrangements and method of product release in the purchasing information.

2.2 Purchasing shall review each FS1090-2 for accuracy and shall sign the form to indicate the review was performed.

2.3 Orders may be placed with the vendor by telephone, fax, mail, or online. When placing orders by telephone, the vendor contact and date of order shall be noted and a confirming copy of the order sent to the vendor.

2.4 Purchasing shall follow up on shipping, delivery, expediting, and partial shipments of ordered items to assist manufacturing with consistent production flow and other departments operational requirements. Purchasing can either telephone vendors or use a FS1090-4 – PURCHASE ORDER FOLLOW-UP to verify, trace or expedite orders.

3.0 RECORDKEEPING AND MATCHING

3.1 When Purchase Orders are issued, the Purchasing and Accounting copies shall be placed in an Open Purchase Order File until the items are received.

3.2 Items shall be received in accordance with FS1110 – Receiving and Inspection. The completed vendor's packing list and/or the Receiving and Inspection Report shall be forwarded to Purchasing.

3.3 Purchasing shall then match the receiving paperwork to the open purchase order. The Accounting copy with the receiving paperwork and supporting requisition paperwork shall be forwarded to Accounts Payable. The Purchasing copy shall be filed in the Closed Purchase Order File.

3.4 For partial shipments, a photocopy of the Purchase Order and the receiving paperwork shall be forwarded to Accounts Payable. The original Purchase Order shall be kept in the open file until all items are received.

4.0 PURCHASING REVIEW

4.1 The Food Safety Team Leader should issue a nonconformity report in the event that a purchased product, upon inspection, fails to meet the requirements for food safety (see FS1110 – RECEIVING AND INSPECTION) or when analysis of FS1110-2 RECEIVING AND INSPECTION REPORTS indicates variations in the supplier's product or suggests a trend away from meeting requirements.

4.2 The Food Safety Team Leader shall periodically audit the purchasing process (annually, at a minimum) to ensure its continuing suitability, adequacy, effectiveness, conformance to various requirements, and ability to deliver safe products to the Company.

Effectiveness Criteria:

- Raw Material Inventory (in days)
- Material availability for Production
- No food safety incidents due to purchased materials, supplies or services

References:

A. Statutory / Regulatory Requirements

There are laws in every country and locality that govern *suppliers* of food products and food ingredients. The Company *purchasing* food products and/or food ingredients has a duty to itself and its customers to ensure suppliers and their products comply with applicable laws. It is the Company's responsibility, therefore, to be aware of the pertinent laws governing food safety practices in every locality where it does business.

B. Food Safety Procedures

- FS1100 - SUPPLIER EVALUATION
- FS1110 - RECEIVING AND INSPECTION

Records:

- FS1090-1 – PURCHASE REQUISITION
- FS1090-2 – PURCHASE ORDER
- FS1090-3 – PURCHASE ORDER LOG
- FS1090-4 – PURCHASE ORDER FOLLOW-UP
- FS1100-1 – APPROVED VENDOR LIST
- FS1110-2 – RECEIVING AND INSPECTION REPORT
- Receiving Documents

Revision History:

Revision	Date	Description of changes	Requested By
0.3	12/30/2005	Initial Release	

FS1090-1 – PURCHASE REQUISITION

Requested by:______________________ Date:______________

Department:______________________ Charge To:______________

Purpose or Use:____

Vendor Name:____

Vendor Address:____

Vendor Address2:____

Vendor Contact:______________________ Phone:______________

Date Needed:______________________ Ship Via:______________

Stock Number	Product / Service Description	Qty.	Unit Price	Extended Cost

For Purchasing Department Use Only

Approvals:

Dept. Manager:__________________Date:__________ Approved:______

Additional:__________________ Date:__________ Ordered for:______

P.O. No.:______

Vendor EIN or SS on file? ❒ Yes ❒ No Date:______

[This page intentionally left blank]

FS1090-2 – PURCHASE ORDER

OUR COMPANY

PURCHASE ORDER

Purchase Order: ________

Order Date: ___________

Page: _______

To:

Ship To:

Ship Via:		Freight Charges:	Terms:		Sales Taxable:
Item No.	**Reference Part No.**	**Product / Service Description**	**Qty.**	**Unit Price**	**Extended Cost**
				Total	

Comments (inspection requirements, etc.):

Authorized Signature:___________________________ Date:___________________

[This page intentionally left blank]

FS1090-3 – PURCHASE ORDER LOG

Purchase Order Number	Date Required	Requesting Department	Engineering drawings and specs	Quantity Required	Recommended Vendor	Special Shipping Requirements

[This page intentionally left blank]

FS1090-4 – PURCHASE ORDER FOLLOW-UP

To: Date:____________________

Please rush a reply to us by fax or telephone on the information requested below. Thank you.

Our Purchase Order #:__________

From: Dated:____________________

Please Respond To Our Request As Indicated Below

❒ Please rush shipment. Advise delivery date:____________________

❒ Has shipment been made? Advise carrier/date:____________________

❒ Partial shipment received. Balance to ship when?____________________

❒ Can you ship in accordance with our requested date? ❒ Yes ❒ No

❒ This shipment shall be shipped via what method?____________________

❒ Price on Terms do not match quotation:____________________

❒ Please review attached and confirm accuracy of all information and prices. Acknowledge below.

❒ These items are not taxable. Our Tax Exempt No. is ____________________. Please revise invoice and acknowledge below.

❒ Incorrect calculations on invoice noted. See attached and verify.

❒ Other____

Comments or Reply:

[This page intentionally left blank]

Doc #: **FS1100**	Title: **SUPPLIER EVALUATION**	Print Date: **2/1/2006**
Rev #: **0.0**	Prepared By:	Date Prepared: **2/1/2006**
Eff. Date: **2/1/2006**	Reviewed By:	Date Reviewed: **2/1/2006**
	Approved By:	Date Approved: **2/1/2006**
Standards: **ISO 22000:2005, clause 7.1**		

Policy: The Company shall do business only with suppliers that provide safe food ingredients / products.

Purpose: To define the criteria and methods for selecting and evaluating suppliers for addition to or disqualification from Our Company's approved supplier list.

Our Company ensures purchased products and services conform to specified requirements. This starts with selection of appropriate suppliers that have the capability and systems to supply products, materials, and services to Our Company's specified requirements, as well as meet relevant statutory/regulatory requirements. Suppliers are controlled to the extent necessary, based on the effect of purchased items on the safety of Our Company's products.

Scope: This procedure applies to all vendors of products, materials, and services that may affect the safety of Our Company's food products.

Responsibilities:

Purchasing is responsible for initial supplier identification and for collection of business information related to the potential supplier. Purchasing is also responsible for maintaining supplier performance data for ongoing evaluation.

Finance is responsible for evaluation of the potential supplier's financial information.

Food Safety Management (FSM) is responsible for evaluating each supplier's food safety systems and for monitoring and reporting on supplier food safety performance on a continuing basis.

Quality Assurance (QA, or Quality) is responsible for evaluating vendors on quality-related issues.

Definitions: Safe food – Food that is not harmful or injurious when consumed; food that does not cause medical illness or pose a health hazard to the consumer.

Supplier (or vendor) – Company/organization that directly supplies Our Company with food; food ingredients; food processing, handling, and/or packaging equipment; and/or other items directly or indirectly related to food safety (e.g., cleaning/sanitation chemicals, labels, containers, equipment maintenance services).

Procedure:

1.0 APPROVED VENDOR LIST

1.1 The Food Safety Team Leader shall maintain an FS1100-1 – APPROVED VENDOR LIST, identifying suppliers who have demonstrated the capability of meeting the company's quality and food safety requirements.

1.2 The FS1100-1 shall be organized alphabetically by product or service supplied and cross-referenced by material ID number and vendor name (or ID).

1.3 In addition to the product or service identification, the FS1100-1 should include, at a minimum:

- Vendor Name;
- Vendor contact information;
- Vendor or Contract number;
- Vendor Class (see section 2.0);
- Vendor's item ID or part number;
- Certification(s); and
- Last audit date.

1.4 To be listed as an approved vendor, a candidate must provide certain assurances of capability, depending on the nature and seriousness of the potential risks its products or services pose to the quality and safety of the company's products.

2.0 VENDOR CLASSIFICATION AND REQUIREMENTS

Vendors shall be classified according to the potential risk their products pose to the safety of the Company's products:

Class I: Vendor's product affects food safety (e.g., raw materials, processing equipment, monitoring/measuring equipment, software).

- Unless already certified to ISO 22000, Class I vendors shall require a second-party audit by a Food Safety Team to the ISO 22000 standard. In the course of the audit, the vendor must provide evidence that all of its processes affecting food safety are effectively documented and implemented. In order to remain on the Company's Approved Vendor List, each vendor must be re-audited at two-year intervals or be certified to the ISO 22000 standard.
- Certification of conformance of a vendor's HACCP plan(s) or PRP(s) may be substituted for the second-party audit requirement, but such

vendors' HACCP plans and PRPs are subject to a second-party audit to verify compliance at any time.

Class II: Vendor's product/service does not affect food safety (e.g., fork lift).

- Verification of initial shipments will be performed on products or services provided by any Class II vendor and vendor performance will be monitored in accordance with Company requirements for non-food-safety-related vendors.

Class III: Vendor's product (service) may or may not affect food safety – the vendor cannot be second-party audited but the Company will grant an exception based on the vendor's performance history (i.e., their QA system is considered sufficient to meet requirements). Examples include express shippers and office supply vendors.

- Verification of *all* shipments will be performed on products and/or services provided by any Class III vendor.

CLASS	ISO Certification or 2nd-Party Audit	Performance Evaluation	HACCP / PRP
Class I	X	X	X
Class II		X	
Class III		X	

Vendor Classification/Requirements Table

3.0 NEW VENDOR EVALUATION

In addition to the requirements for the approved vendor list for each class of vendor listed above, Food Safety Management shall evaluate new vendors using the following criteria:

3.1 The vendor's performance capability (i.e., financial status, sufficient facilities, capability of equipment, and capability/training of employees), ability to fulfill Company requirements, and ability to deliver accurately, completely, and in a timely manner. NOTE: Food Safety Management shall fill out an FS1100-2 – VENDOR SURVEY FORM for all prospective Class I vendors.

- Competitive pricing is but one component of the evaluation; pricing will not be the deciding factor unless competing vendors are equal in every other area.
- Prospective vendors shall be subject to reference checks.

3.2 Vendors certified to the ISO 22000:2005 standard shall be given preference.

- When such vendors cannot be found, the Company shall give preference to vendors having HACCP-based plans and PRPs in place. The Company shall verify the prospective vendor's HACCP plans and/or PRPs.
- A copy of the vendor's ISO certificate and/or verification of vendor HACCP plans and PRPs shall be kept on file.

3.3 When Food Safety Management adds a vendor to the approved vendor list, it shall notify Purchasing, Accounting, and the likely requisitioning department using form FS1100-3 – APPROVED VENDOR NOTIFICATION.

4.0 VENDOR DISQUALIFICATION

4.1 A supplier (vendor) may be disqualified, based on the degree of nonconformity in a single incident, in accordance with FS1150 – CONTROL OF NONCONFORMING PRODUCT and FS1170 – CORRECTIVE ACTION.

4.2 At periodic reevaluation, FS1100-4 – VENDOR PERFORMANCE LOG may indicate a vendor's safety performance rating in the period being evaluated is unacceptable.

4.3 Disqualified vendors should be removed from the Approved Vendor List. Alternately, there should be a "qualified / disqualified" indicator on the Approved Vendor List.

4.4 The date of disqualification shall be noted.

- A disqualified vendor may be reinstated no sooner than 90 days following a disqualification, subject to reevaluation.

5.0 VENDOR REEVALUATION

5.1 Food Safety Management, with the assistance of Quality Assurance, shall periodically evaluate the performance of each vendor on the Approved Vendor List (annually, at a minimum) for the following factors:

- History of supplying safe products or services (i.e., safety performance rating).
- Quality rating = Items (lots) rejected ÷ Total items (lots) received x 100. Ratings less than 95% require corrective action (see FS1170 – CORRECTIVE ACTION). Exceptions to the 95% Corrective Action requirement may be given where the total quantity of items or lots received is small and at the Quality Manger's discretion.
- Responsiveness to Corrective Action Requests or ability to correct problems with delivery or quality.
- Second-party audit results.
- Status of certification (ISO, HACCP plan).
- On-Time Delivery, 100% on time expected (0 days early, 0 days late).

5.2 FSM, with the help of QA, shall reevaluate disqualified vendors on the same factors that led to their disqualification (see section 4.0).

5.3 A record of each vendor shall be maintained on a separate form FS1100-4 – VENDOR PERFORMANCE LOG for purchasing use. Information on the Vendor Performance Log shall include, at a minimum:

- Vendor Name & contact info
- Vendor Class
- Certification
- Last audit date, auditing authority
- Last safety inspection, inspecting authority
- Quality/Food Safety contact
- Order date
- Required / actual delivery date
- Quantity ordered / delivered
- Deviation (discrepancy)
- NCR (nonconformity report)
- CAR status

6.0 SECOND-PARTY VENDOR AUDIT PROGRAM

6.1 Food Safety Management shall audit the Company's Class I vendors on a periodic basis. See FS1160 – INTERNAL AUDIT AND SYSTEM VALIDATION for guidance on conducting the audit.

6.2 Top Management shall reevaluate the second-party vendor audit program periodically, to ensure that Class I vendors are being audited on a timely basis.

Effectiveness Criteria:

- Number of approved suppliers
- Average Supplier Quality Ratings
- Average On-time Delivery ratings

References:

A. Statutory / Regulatory Requirements

Laws in every country and locality govern *suppliers* of food products and food ingredients. The Company *purchasing* food products and/or food ingredients has a duty to itself and its customers to ensure suppliers and their products are complying with applicable laws and providing safe products. Furthermore, many governments require traceability of products throughout the supply chain – companies are being required to implement "one-up / one-back" traceability, where they document all products from their suppliers and to their customers.

It is the Company's responsibility to be aware of the pertinent laws governing food safety practices in every locality in which it does business.

B. Food Safety Procedures

- FS1050 – PREREQUISITE PROGRAMS
- FS1080 – HACCP PLAN MANAGEMENT
- FS1160 – INTERNAL AUDIT AND SYSTEM VALIDATION
- FS1170 – CORRECTIVE ACTION

Records:

- FS1100-1 – APPROVED VENDOR LIST
- FS1100-2 – VENDOR SURVEY FORM
- FS1100-3 – APPROVED VENDOR NOTIFICATION
- FS1100-4 – VENDOR PERFORMANCE LOG
- Vendor corrective actions
- Vendor audit results

Revision History:

Revision	Date	Description of changes	Requested By
0.0	2/1/2006	Initial Release	

FS1100-1 – APPROVED VENDOR LIST

Item ID Number	Item Description	Vendor Name/ Contact Info	Vendor or Contract Number	Vendor Class (I-II-III)	Vendor Item ID (part) Number	Vendor Cert. ID and Cert. Body	Cert. Eff. Until (date)	Last Audit Date	Notes

[This page intentionally left blank]

FS1100-2 – VENDOR SURVEY FORM

Date:____________________

Vendor Name: ____

Address: ____

____

____

Phone : (______) ______ Fax: (_____) ____________

Quality Contact:________Title:____________________

Phone : (______) ______Fax: (______________) __________

Food Safety Contact: :____________________Title:____________________

Phone : (______) ______ Fax: (______________) __________

PART I - GENERAL INFORMATION

Product for which survey was performed ________________________________

List Company Management

Name	Title	Interviewed

List at least two credit references

Company Name	Account or Reference number	Phone

List at least two trade references

Company Name	Account or Reference number	Phone

Number of years in business: ____________________

Certified? ☐ ISO 22000:2005 ☐ HACCP plan (describe; attach documents, if needed)

Date of last audit: ___/___/_____

Name of Registrar: ____

TYPE OF OPERATION

Retail ☐
Distribution ☐
Manufacturing/Processing ☐
Crop/Animal/Seafood Production ☐
Other ☐

BUSINESS ACTIVITIES

Purchasing ☐
Warehousing ☐
Design ☐
Research/Development ☐
Production ☐
Manufacturing ☐
Sales/Customer Service ☐
Retail ☐
Distribution ☐
Training ☐
Consulting/Business Services ☐
Other ☐
(Describe) ____________________

____

____

PART II - RAW MATERIALS

PURCHASING

Is Qualification Based On Written Specifications And Approval Of Vendor Sources?
☐ Yes ☐ No

Are Reject/Accept Limits Shown?
☐ Yes ☐ No

Is Approval Based On:
☐ Quality History ☐ Supplier ☐ On-Site
☐ Own QC ☐ Cards ☐ Survey
☐ Certificate ☐ Testing
☐ Other____________________

Are Specification Changes Reviewed And Signed Off By QC Personnel?
☐ Yes ☐ No

TESTING

Are Written Test Procedures In Use?
☐ Yes ☐ No

Are Test Results On File?
☐ Yes ☐ No

Is a Sampling Plan Used?
☐ 100% ☐ Mil Spec ☐ AQL ☐ Random
☐ Other____________________

Do Test Results Indicate
☐ Quantity Sampled
☐ Method Of Analysis
☐ Date/Signature Of Analyst
☐ Sample Traceability

Is There A Retention Sample System For Raw Materials/Components?
☐ Yes ☐ No

IN PLANT CONTROL

Is Material Assigned Alpha-Numeric Or Identifying Mark For Each Incoming Lot?
☐ Yes ☐ No

Is Material Visibly Marked As
☐ Sampled ☐ Approved
☐ Rejected ☐ Not Marked

Is an Inventory Log Or Record Kept?
☐ Yes ☐ No

Is Storage Area Separate?
☐ Yes ☐ No

Is Storage Area Segregated?
☐ Yes ☐ No

Is a Stock Rotation (FIFO) System Used?
☐ Yes ☐ No

Is There Authorized Custodian Control?
☐ Yes ☐ No

Is General Housekeeping Neat And Orderly? ☐ Yes ☐ No

Rejected Materials Are:
Clearly Identified ☐ Yes ☐ No
Physically Segregated ☐ Yes ☐ No

PART III - MANUFACTURING

MASTER PRODUCTION RECORDS

Is there a single controlled file of master records for each product?

☐ Yes ☐ No

Are these master records signed and dated?

☐ Yes ☐ No

Double signature ☐ Yes ☐ No

Revision dates ☐ Yes ☐ No

Are the process, assembly, or manufacturing steps fully described:

In the master production record?

☐ Yes ☐ No

In a separate document or record?

☐ Yes ☐ No

Does the master document indicate:

QC points for in-process mfg?

☐ Yes ☐ No

Type of test or inspection to be made?

☐ Yes ☐ No

Method of measurement?

☐ Yes ☐ No

Who performs test or inspection?

☐ Yes ☐ No

Level of accept/reject (limits)?

☐ Yes ☐ No

For manufacturing, processing, subassembly, or packaging done by outside sources, are there:

Master production records?

☐ Yes ☐ No

QC specifications and methods records?

☐ Yes ☐ No

Outside sources not used?

☐ Yes ☐ No

PRODUCTION AREA

Is the work flow organized?

☐ yes ☐ no

Distinct staging area for raw materials or Components used in manufacturing?

☐ yes ☐ no

Production or assembly lines segregated?

☐ yes ☐ no

Are general housekeeping and environmental factors adequate?

☐ yes ☐ no

Are written procedures for plant sanitation available?

☐ yes ☐ no

PRODUCTION EQUIPMENT

Are maintenance or service records available?

☐ yes ☐ no

Is calibration performed on a periodic basis?

☐ yes ☐ no

Are there means of readily identifying type and stage of processing being done on equipment?

☐ yes ☐ no

PRODUCTION RECORDS

Are production documents collected and filed?

☐ yes ☐ no

Production documents kept ______(years)

☐ complete history

☐ labeling samples included

☐ partial history

☐ traceability by lot or serial#

PACKAGING

Are finished goods packaging operations segregated?

☐ Yes ☐ No

Finished goods under supervised control?

☐ Yes ☐ No

Label records kept?

☐ Yes ☐ No

Pre-Label:

☐Count ☐ Reconciliation

Are finished goods properly identified, labeled, and stored?

☐ Yes ☐ No

☐ prior to release ☐ After release

REJECTED MATERIALS

Are there written procedures for disposing of or reworking rejected items?

☐ Yes ☐ No

Are rejected products held in quarantine pending final disposal?

☐ Yes ☐ No

☐ Held in segregated area

☐ With special markings

RETENTION SAMPLES

Are samples of finished goods retained?

☐ Yes ☐ No

☐ From each production run

☐ In a separate controlled area

☐ In the same container/closure system in which they are sold

☐ In containers different from unit as sold

☐ kept for a period of ______ (years)

☐ Written log or file

STERILE COMPONENTS (If Applicable)

Are there procedures for establishing and maintaining aseptic conditions?

☐ Yes ☐ No

Are there methods for routine auditing of sterile areas used?

☐ Yes ☐ No

Are there procedures for working in sterile areas?

☐ Yes ☐ No

For cleaning and sterilization of equipment?

☐ Yes ☐ No

For bulk and final product sterility testing?

☐ Yes ☐ No

Is process sterility for each run documented in the production records?

☐ Yes ☐ No

Are sterile processes used?

☐ Radiation ☐ Steam ☐ ETO

☐ Filtration ☐ Chemical

Other:______________________________

____

PART IV - QUALITY CONTROL / ASSURANCE

ORGANIZATION AND FUNCTION

Does the quality control-inspection group report directly to the top, independent of production, marketing, or other organization groups within the manufacturing company?

☐ Yes ☐ No

Does the quality control-inspection group have full authority to withhold shipment or further production of rejected items?

☐ Yes ☐ No

ORGANIZATION AND FUNCTION (Continued)

Are the quality control procedures revised on a periodic basis?

☐ Yes ☐ No

Does the quality control/assurance-inspection group have:

Education, training, or experience?

☐ Yes ☐ No

Understanding of their function?

☐ Yes ☐ No

OPERATIONS

Are stamps, tags, markers, etc. used to verify inspection activity?

☐ Yes ☐ No

Are the markings used traceable to an individual inspector?

☐ Yes ☐ No

PART V - CUSTOMER COMPLAINTS AND RECALL CAPABILITIES

Is there an organized complaint file system?

☐ Yes ☐ No

Does each complaint state:

- ☐ Nature of complaint
- ☐ Response to customer (repair, refund, replace)
- ☐ Further corrective action by manufacturer

Complaint files kept for _______ (years)

Is there a periodic review of complaint files for trends? ☐ Yes ☐ No

Is the review file as a written summary?

☐ Yes ☐ No

Is there a group or individual assigned to handle customer inquiries and follow up on complaints?

☐ Yes ☐ No

Are product defects verified by manufacturer through testing?

☐ Yes ☐ No

Was review of complaint files for survey product made?

☐ Yes ☐ No

Are production samples for qc testing: adequately identified as to source

☐ Yes ☐ No

recorded somewhere at time of sampling?

☐ Yes ☐ No

entered on filed test report?

☐ Yes ☐ No

Written sampling plan based on:

☐ 100% ☐ Mil. Spec ☐ AQL ☐ Random

Other___________________________

Is the product used tested prior to final release?

☐ Yes ☐ No

Are outside sources used for production testing?

☐ Yes ☐ No

- ☐ Under formal contract
- ☐ Used test protocols
- ☐ Written procedures
- ☐ Copies in the manufacturing file
- ☐ Facility registered or licensed by any federal, state or professional agency
- ☐ Outside test results filed by manufacturer

Is there a formal quality assurance program involving performance testing of the product(s) after release?

☐ Yes ☐ No

RECALL CAPABILITIES

Is there a company recall plan?

☐ Yes ☐ No

☐ Shows how decisions are made and by whom

☐ How recall will be accomplished

☐ Instructions for recovery and accountability of recalled product

Do shipping or distribution records on file show:

Customer/distributor name and address?

☐ Yes ☐ No

Date of shipments and quantity shipped?

☐ Yes ☐ No

Lot or serial number of product?

☐ Yes ☐ No

Distribution records are maintained ____ (years)

Distribution records are stored as:

☐ Computer listing

☐ Microfilm/microfiche

☐ Manual card/paper files

PART VI – REGULATORY COMPLIANCE

Are all necessary approvals for marketing products available?

☐ Yes ☐ No

Is there a file with past and current labeling for each survey product?

☐ Yes ☐ No

Is there a formal auditing program of the food safety operation? (If so, done by whom?)

☐ Yes ☐ No

____

____

____

FOOD SAFETY

Does the vendor have HACCP plans in place, as required?

☐ Yes ☐ No

Does the vendor have PRPs in place?

☐ Yes ☐ No

(If "yes" to either of above, list and attach to this document.)

List of Attachments and Comments

____

____

____

____

NOTICE: I/we certify that the information contained in the attached survey form is accurate and complete as of the date indicated. Where trade secret or other proprietary information is involved, the person interviewed has initialed those responses not verified by the interviewer. All information obtained will be kept confidential. A corporate officer of the Company surveyed will review all responses made at the time of survey. This survey has been made with the permission of the Company surveyed

Signature	Title	Location
Signature	Title	Location

FS1100-3 – APPROVED VENDOR NOTIFICATION

Date:_______________

Product for which approved: ______________________________

Vendor Name: ______________________________

Address: ______________________________

____

____

Sales Contact Person: ______________________________

Phone : (   ) ______________ Fax: (   ) ______________

Quality Contact: ______________ Title: ______________

Phone: ______________

Food Safety Contact: ______________ Title: ______________

Phone: ______________

Class: ______________ Certification: ______________

Registrar: ______________

Prepared by (source): ______________

Approved by: ______________

Distribution: Purchasing (Vendor File)
Accounting
Requisitioning Department ______________

[This page intentionally left blank]

FS1100-4 – VENDOR PERFORMANCE LOG

Vendor Name ______________ Quality/Food Safety Contact ______________

Address ______________ Q/FS Phone Number ______________

Address ______________ E-Mail ______________

Vendor Phone ______________ Vendor Class ☐ I ☐ II ☐ III

Product ______________ Certification ☐ ISO 22000[1] ☐ HACCP plan (specify[2])

Product ______________

Last Audit Date ______________ Auditing Authority ______________

Last Safety Inspection ______________ Inspecting Authority ______________

Product	Order Date	Quantity Ordered	Quantity Received	Back Order Qty	Requested Delivery Date	Actual Delivery Date	Deviation (Discrep-ancy)	Order Cycle Time	Shipment Damaged Y/N	Shipment Rejected Y/N	NCR Number	CAR Status

[1] See FS1100-1 – APPROVED VENDOR LIST for certification ID, etc.
[2] Attach additional sheets, if necessary.

Product	Order Date	Quantity Ordered	Quantity Received	Back Order Qty	Requested Delivery Date	Actual Delivery Date	Deviation (Discrep-ancy)	Order Cycle Time	Shipment Damaged Y/N	Shipment Rejected Y/N	NCR Number	CAR Status

Doc #: **FS1110**	Title: **RECEIVING AND INSPECTION**	Print Date: **2/1/2006**
Rev #: **0.0**	Prepared By:	Date Prepared: **2/1/2006**
Eff. Date: **2/1/2006**	Reviewed By:	Date Reviewed: **2/1/2006**
	Approved By:	Date Approved: **2/1/2006**
Standards: **ISO 22000:2005, clause 7.1**		

Policy: The Company shall ensure food safety by inspecting incoming ingredients or products that may have an effect on the safety of its products.

Purpose: This procedure outlines the steps for the receiving and inspection of materials, components, parts, finished goods, etc., and the stocking of these items or the disposition of rejected items.

Scope: This procedure applies to the receipt of all inventory items.

Responsibilities:

Receiving and warehouse personnel are responsible for receiving materials, inspecting incoming shipments for completeness and integrity (and rejecting out-of-compliance shipments), and forwarding all paperwork to Purchasing for payment.

Accounting and Accounts Payable are responsible for payment of invoices only after satisfactory completion or delivery of goods or services has been made.

The Food Safety Team (Food Safety) is responsible for reviewing all rejections by the Food Inspector, completing Part III of the Receiving and Inspection Report, and forwarding the R&IP to Purchasing.

The Food Inspector is responsible for inspecting and authorizing or rejecting incoming food and food ingredients on the basis of food safety.

Procedure:

1.0 RECEIVING

1.1 Receiving shall visually examine all incoming shipments for apparent package damage promptly. If the shipment has apparent damage, Receiving shall notify the Food Safety Team immediately. Food Safety shall decide to either accept or reject the shipment from the carrier. Damaged incoming shipments should cause a nonconformity report to be issued. See FS1170 – CORRECTIVE ACTION.

1.2 If the shipment shows no signs of damage or Food Safety has decided to accept a damaged shipment, Receiving shall count the pieces (i.e., packages, skids, barrels) and confirm with the bill of lading and note any exceptions (i.e., package damage or shortages).

1.3 Receiving shall obtain the packing list from the shipment and record the appropriate information in the FS1110-1 – RECEIVING LOG. Receiving shall also record the date and log number on the vendor's packing list.

1.4 Receiving shall deliver the incoming shipment with its packing list (or Receiving and Inspection Report) to the appropriate area for inspection.

2.0 INSPECTION

2.1 At the appropriate receiving area, Receiving shall unpack each shipment, count all items, and match items to the packing list. Receiving should complete part I of the FS1110-2 – RECEIVING AND INSPECTION REPORT.

2.2 Receiving shall inspect the shipment for conformance to specific inspection requirements listed on the PO and for conformance to Company requirements. If different materials are included in the shipment, each will be segregated and inspected accordingly. Receiving shall complete part II of the FS1110-2.

2.3 The Food Inspector shall conduct inspections of incoming food and food ingredients in accordance with controlling federal, state, and local regulations.

3.0 REJECTION, DISCREPANCIES, AND DISPOSITION

3.1 Any quantity discrepancies will be noted on the packing list or Receiving and Inspection Report, which shall be signed and forwarded to Purchasing. Purchasing shall follow up with the vendor to resolve the shipping discrepancy.

3.2 If there is a nonconformity other than quantity (e.g., broken outer carton), the suspect goods shall be handled in accordance with FS1150 – CONTROL OF NONCONFORMING PRODUCT. Goods rejected by the Inspector shall be disposed of in accordance with controlling statutes/regulations.

Food Safety shall review all rejections, complete Part III of FS1110-2 – RECEIVING AND INSPECTION REPORT, and forward the FS1110-2 with the packing list to Purchasing.

3.3 Suppliers shall be evaluated within twenty-four (24) hours of when discrepancies, nonconformities, and/or rejections are found, in accordance with FS1100 – SUPPLIER EVALUATION.

4.0 STOCKING

4.1 Goods found to be in conformance or suitable for their intended use and accepted by the Food Inspector shall be moved to the appropriate storage area by Receiving. Food Safety may note on the FS1110-2 report the justification for any materials accepted and shall forward the FS1110-2 with the packing list to Purchasing.

4.2 Accepted materials shall be clearly labeled or otherwise identified in accordance with FS1050 – PREREQUISITE PROGRAMS.

- Identification shall indicate the material designation, PO number, quantity, date received, expiration date, and any special preservation requirements, in accordance with FS1130 – IDENTIFICATION, LABELING, AND TRACEABILITY.

5.0 RECEIVING AND INSPECTION REVIEW

Incoming goods arriving less than 30 minutes prior to the end of the shift may be stored in receiving as long as environmental storage requirements are met. Items arriving more than 30 minutes prior to the end of the shift must be processed and moved to storage before the end of the shift. Exceptions to this procedure should be reported via a Nonconformity Report (NCR).

Effectiveness Criteria:

- Timely processing of incoming goods
- Timely inspection of received goods (cycle time)
- Material availability for Production.
- No nonconforming (contaminated) material introduced into the production process
- No errors in receiving records

References:

A. Food Safety Manual

- FSM 7.1 – Planning and Realization of Safe Products-General

B. Regulatory / Statutory Requirements

The Company must be familiar with and comply with the controlling food safety laws in the various places where they do business. Some examples of laws pertaining to inspection include: the Fish Inspection Act (Canada); Title 9, Chapter 3, section 417 of the Code of Federal Regulations (USA); and Regulation (EC) No 854/2004 of the European Parliament (EU).

Regulations regarding the scope, frequency, and nature of inspections conducted by government inspectors (CFIA, FSIS, etc.) will vary according to the nature of the food and/or food ingredients being inspected and the potential for risks to the consumer.

C. Food Safety Procedures

- FS1090 – PURCHASING
- FS1100 – SUPPLIER EVALUATION
- FS1130 – IDENTIFICATION, LABELING, AND TRACEABILITY

Records:

- FS1090-1 – PURCHASE REQUISITION
- FS1090-2 – PURCHASE ORDER
- FS1110-1 – RECEIVING LOG
- FS1110-2 – RECEIVING AND INSPECTION REPORT
- FS1130-1 – LOT IDENTIFICATION / TRACEABILITY LOG
- FS1150-1 – NONCONFORMANCE REPORT

Revision History:

Revision	Date	Description of changes	Requested By
0.0	2/1/2006	Initial Release	

FS1110-1 – RECEIVING LOG

Log #	Date-Time Rcvd.	Date/Time Moved to Storage	Supplier	Item Desc.	Supplier Item ID	Supplier Lot #	Carrier	Carrier Tracking #	Our Purchase Order #	Qty. OK Y/N	Damage Y/N	Rcvg. Initials

[This page intentionally left blank.]

FS1110-2 – RECEIVING AND INSPECTION REPORT

PART I – RECEIVING

Date: ______________

Vendor/Supplier/Subcontractor: ______________________________

P.O. Or Contract No.: ______________________________

Item No.	Description	INSPECTION Criteria	QUANTITIES					
			Ordered	Received	Inspected	Accepted	Rejected

Received By: ______________________

PART II – INSPECTION

Inspected By: ______________________

Sample Lot

Conformance/Discrepancies to Specifications

	YES	NO		YES	NO
Shipping Damage	❒	❒	Functional	❒	❒
Markings/Finish	❒	❒	Dimensions	❒	❒
Attributes	❒	❒	Other	❒	❒

Lot Size: __________

Sample Qty: ________

Accepted: ____________ Date: _________ Rejected: ________ Date: _____

❒ Place in Stock

❒ Forward to Next Operation

____

____

Cause for Rejection: ______________________

____

____

____

PART III – REJECTED MATERIAL DISPOSITION

Return to Vendor

____

Signature

Conditional Acceptance Approvals

____

Signature

Remarks: ______________________

____

Further comments may be noted on back of report or additional sheets if necessary.

[This page intentionally left blank]

Document ID: **FS1120**	Title: **MANUFACTURING**	Print Date: **2/1/2006**
Revision #: **0.0**	Prepared By:	Date Prepared: **2/1/2006**
Effective Date: **2/1/2006**	Reviewed By:	Date Reviewed: **2/1/2006**
	Approved By:	Date Approved: **2/1/2006**
Standards: **ISO 22000:2005, clauses 7.1, 7.6.4, 8.1, and 8.2**		

Policy: The Company shall develop, implement, and maintain product realization processes that result in safe products.

Purpose: To delineate the processes used to manufacture, inspect, package, and release safe product for customer use.

Scope: This procedure applies to all personnel involved in the handling, production, and inspection of Company products and ingredients.

Responsibilities:

Warehouse personnel are responsible for picking product materials (ingredients, additives, etc.) for production.

The Food Safety Team is responsible for reviewing revisions to manufacturing practices, to ensure potential food safety hazards have been adequately addressed.

The Production Manager (Supervisor) is responsible for managing the Company's production process(es), ensuring that all personnel, materials, and equipment needed to create and maintain the production schedule are available and able to function as expected. The Production Manager is also responsible for ensuring that Production personnel receive adequate training to perform their assigned duties.

Manufacturing (production) personnel are responsible for producing the Company's end product(s) in a manner that ensures food safety and meets statutory/regulatory and customer requirements.

Quality Control is responsible for monitoring and measuring all manufacturing (product realization) processes, to ensure Company products conform to food safety requirements.

Definitions: Product realization – The act of bringing a product into existence, or simply "making a product". (ISO came up with the term to include *services* in the product realm; in the context of this procedure, "product realization", "manufacturing", and "production" are synonymous.)

Good manufacturing practice – A practice (or set of practices) designed to ensure that food products are manufactured and tested according to prescribed food safety standards.

Flow diagram – Paths of movement (of workers / materials) superimposed on a graphical representation of the work area. A flow diagram may also be called a "process map".

Procedure:

1.0 KITTING WORK ORDERS

1.1 At the start of each shift, the Production Supervisor shall review the Production Schedule and work orders delivered by Production Scheduling to ensure adequate personnel and equipment capacity to meet due dates.

1.2 The Production Supervisor shall issue pick lists to the warehouse (storage) to kit the work orders. The pick list indicates part numbers and locations for the materials (ingredients) to be issued to the work order.

1.3 The assigned material handler shall pull the indicated items and record the quantity and location pulled from on the pick list. The handler should mark the pulled package/container with the work order number(s) to which it is being issued, especially if those items are being pulled for more than one work order at the same time and location.

1.4 If ingredient packages/containers must be broken to issue the correct quantity, the material handler shall place the correct quantity into an appropriate container and clearly label the container with the ingredient name, quantity, purchase order number, and work order number.

- If the *appropriate quantity* of an ingredient is not in the location indicated on the pick list, the material handler shall attempt to find additional locations for the ingredient on the Company inventory system and, if found, note the quantity and actual location on the pick list.
- If the ingredient *cannot be located in the plant*, the material handler shall contact the Production Supervisor immediately to determine if the kit should be held in storage until complete and/or the work order should be rescheduled. The Production Supervisor should immediately initiate corrective action, in accordance with FS1170 – CORRECTIVE ACTION.

1.5 The material handler shall deliver the completed kit to Production with a copy of the completed pick list.

1.6 Using the original pick list, Production shall enter all ingredient issues from the actual locations on the computer system. If location discrepancies were noted during kitting, Production shall notify Material Control.

2.0 PRODUCTION

2.1 The Production Supervisor shall assign the work order for the first operation on the Router (routing sheet).

2.2 The Operator shall move the ingredients issued on the work order to the work area.

2.3 The Operator shall follow the operating instructions on the Router. The Router may indicate or reference more detailed work instructions (CCP, critical limits, etc.).

2.4 When the operation is complete and the operator is satisfied all product criteria are acceptable, the Operator shall notify Food Safety for an inspection. The product cannot be released for further processing, packaging, storage, etc., until the safety inspection is complete and the Inspector records the results and stamps the inspection sheet.

- If the product fails any inspection criterion, the Inspector shall handle the product in accordance with FS1150 – CONTROL OF NONCONFORMING PRODUCT.

2.5 The Operator shall sign the inspection sheet and complete the Router for the operation. The quantity accepted is indicated on the router and initialed by the Operator.

2.6 When the operation is complete, the Operator shall forward the product and work order packet to the next operation on the Router.

3.0 FINAL INSPECTION

3.1 The last operation before Packaging on every Router is Final Inspection.

3.2 Food Safety performs any final inspections as required on the Inspection Sheet.

- Pass = Record results on the Inspection Sheet and stamp the Router and Inspection Sheet.
- Fail = Handle product according to FS1150 – CONTROL OF NONCONFORMING PRODUCT.

4.0 PACKAGING AND LABELING

4.1 Food Safety prints the appropriate quantity of labels, with lot/serial numbers and other required information, after the product has passed Final Inspection.

4.2 The packaging operator applies any labels required to products after Final Inspection.

4.3 ne operator records serial number sequence on the Router, if applicable.

4.4 The operator packages finished products according to work instructions for the product.

4.5 The packaging operator applies the box labels to the packaged product and stages the product for release in the appropriate area.

4.6 If the product is an ingredient to be used in a subsequent operation, the operator places the product in an appropriate container and attaches a tag, indicating the ingredient name/ID and work order number on the tag.

5.0 FINAL RELEASE

5.1 Food Safety shall review all documentation in the work order packet to ensure it is complete.

5.2 Food Safety shall verify that all inspections have been performed and the product has passed all inspections.

- If reworked nonconforming product has been included in the batch to be released, Food Safety shall verify that reworked product inspections have been performed and that the reworked product has passed all inspections.

5.3 Food Safety shall apply a food safety stamp to each end product box label (or ingredient container) and stamp the Release section of the Router and inspection sheet.

5.4 Storage personnel shall move end product to the appropriate area for shipping.

Effectiveness Criteria:

- Product(s) completed by due date(s).
- Consistency in meeting product due dates.
- Low reject and waste disposal rates.
- Elimination of food safety hazards or reduction to acceptable levels.
- Reduction, elimination, or prevention of product recalls.

Additional Resources:

A. Current Good Manufacturing Practice in Manufacturing, Packing, or Holding Human Food, 21 CFR 110, U.S. Food and Drug Administration – see http://www.fda.gov.

B. A Guide to Good Manufacturing Practice, 2nd Ed., Drug & Market Development Publishing (Nov., 2004).

C. A WHO Guide to Good Manufacturing Practice Requirements, World Health Organization, Geneva, Switzerland (1997).

References:

A. Food Safety Manual:

- FSM 7.5.4 – System for Monitoring Critical Control Points
- FSM 8.1 – Validation, Verification, and Improvement of the FSMS – General
- FSM 8.2 – Validation of Control Measure Combinations

B. Food Safety Procedures:

- FS1000 – DOCUMENT CONTROL
- FS1010 – FOOD SAFETY RECORDS

- FS1080 – HACCP PLAN MANAGEMENT
- FS1150 – CONTROL OF NONCONFORMING PRODUCT

C. Statutory / Regulatory Requirements

Around the globe, there are many statutes/regulations that pertain to the production of safe food or food ingredients. In the USA, the FDA is responsible for enforcing 21 CFR 110, "Current Good Manufacturing Practice in Manufacturing, Packing, or Holding Human Food". Also, the USDA FSIS (US Department of Agriculture Food Safety and Inspection Service) is responsible for enforcing federal regulations pertaining to meat and poultry.

The Company is required to be aware of pertinent food safety statutes/regulations and conduct its operations in accordance with all applicable laws.

Records:

- Production Schedule
- Pick List
- Shortage Report
- Completed work orders and routers
- Completed Inspection Sheets
- FS1150-1 – NONCONFORMANCE REPORT
- Food labeling (labels for attaching to individual packages or cartons; preprinted packaging; etc.)

Revision History:

Revision	Date	Description of changes	Requested By
0.0	12/30/2005	Initial Release	

[This page intentionally left blank.]

Doc #: **FS1130**	Title: **IDENTIFICATION, LABELING, AND TRACEABILITY**	Print Date: **2/1/2006**
Rev #: **0.0**	Prepared By:	Date Prepared: **2/1/2006**
Effective Date: **2/1/2006**	Reviewed By:	Date Reviewed: **2/1/2006**
	Approved By:	Date Approved: **2/1/2006**
Standards: **ISO 22000:2005, clauses 7.3.3.2 and 7.9**		

Policy: The Company shall keep sufficient records of all food safety-related processes to ensure proper disclosure of product information and ensure product / ingredient traceability.

Purpose: To outline the content and format requirements for identification and caution labels attached to products.

Scope: This procedure applies to all food products manufactured and sold by the Company.

Responsibilities:

Receiving is responsible for identifying raw materials / ingredients as they are delivered to the Company, confirming completeness and initial acceptability of incoming orders, notifying Food Safety Inspection when orders are ready for inspection, and for moving inspected items to their appropriate storage areas.

Food Safety Inspection (which may consist of an internal Food Safety Team or one or more certified, government-authorized inspectors) is responsible for updating raw materials inventory with inspection results.

Manufacturing (processing) is responsible for maintaining the Company's raw materials inventory records as foods and/or ingredients are introduced to the production process and for maintaining an intradepartmental log of items used (descriptions, sources, destinations, etc.).

Packaging/Storage is responsible for logging finished products as they are moved from processing to storage.

Shipping is responsible for maintaining a log of all information on finished products going to the Company's customers.

Definitions: End product – Product that will undergo no further processing / transformation within the Company.

Raw material – Material the Company receives from suppliers, then processes (transforms), with or without other raw materials, into an end product.

Traceability – Ability to trace history, application, or location of something (document, work product, ingredient, etc.) by means of recorded information.

Procedure:

1.0 LABELING AND TRACEABILITY – GENERAL

1.1 Recordkeeping is one of the foundation principles of the Company's HACCP plan. When done correctly, Company records will show that:

- Correct procedures have been followed from start to finish;
- All materials used in production are accounted for, so that labeling is complete and accurate, as required;
- All products are traceable, internally and externally;
- The Company conforms to requirements; and
- The Company is capable of responding promptly and effectively in the event of a product recall.

1.2 Within the production chain, the Company should be able to trace materials at two levels:

- Internal traceability, where raw materials and internal processing are traceable within the Company to the final product; and
- Chain traceability, where information accompanies a product from one link in the food supply chain to the next, extending traceability of any product to every link in the supply chain.

2.0 LABELING / TRACEABILITY SYSTEM DEVELOPMENT

2.1 The Food Safety Team Leader, in conjunction with the QA Manager, shall develop and oversee implementation of a system that enables identification of product lots and their relation to batches of raw materials, processing, and delivery records.

2.2 The Company's traceability system shall enable identification of incoming material from immediate suppliers and the initial distribution of its end product.

2.3 The Company shall maintain traceability records for a period defined by law and dependent on the nature of the food / ingredient, to enable the handling of potentially unsafe products and in the event of product recall / withdrawal (see FS1200 – PRODUCT RECALL and FS1010 – FOOD SAFETY RECORDS).

3.0 LABELING / TRACING FOOD ITEMS AND INGREDIENTS

3.1 Receiving shall record information on incoming food items / ingredients in accordance with FS1110 – RECEIVING AND INSPECTION.

3.2 Food Safety Inspection shall maintain records on all items tested for traceability purposes, in accordance with FS1010 – FOOD SAFETY RECORDS. Such inspection records should include, but need not be limited to:

- Description of inspected item(s);
- Lot or group number;
- Date / time inspected;
- Inspection result(s); and
- Special instructions.

3.3 Food / ingredients that do not pass inspection shall be handled in accordance with FS1150 – CONTROL OF NONCONFORMING PRODUCT.

3.4 Production (processing) department managers shall appoint production / operations employees to maintain lot identification / product traceability logs for every product line or item the Company produces. FS1130-1 – LOT IDENTIFICATION / PRODUCT TRACEABILITY LOG EXAMPLE should be used for guidance.

3.5 Employees responsible for *packing* (packaging) and storing food products shall ensure product containers are clearly labeled with the following information, at a minimum:

- Brand name and/or generic name;
- List of ingredients, as required by law;
- Handling and storage requirements;
- Expiration date;
- Lot or group number(s);
- Quantity; and
- Packing date / time.

Additional hazard / health information (e.g., possible presence of food allergens) shall be listed as required by applicable statutes / regulations.

3.6 At a minimum, Shipping shall record the following information (in addition to the requirements in 3.5) upon releasing products to the shipper/customer:

- Name of firm receiving product;
- Name of individual responsible for receiving product;
- Receiver's location / address;
- Contact information (phone / fax number, e-mail address, etc.); and
- Type of food (ingredient), including brand name and specific variety.

3.7 All food / food ingredient information shall be maintained in accordance with FS1010 – FOOD SAFETY RECORDS.

4.0 LABELING / TRACEABILITY SYSTEM REVIEW

4.1 The QA Manager should periodically[1] review logs and other records generated in the course of product processes to determine if the labeling and traceability system is being properly implemented and continues to conform to Company and statutory / regulatory requirements. The QA Manager shall report findings and observations to the Food Safety Team Leader for review.

4.2 A third-party audit of the labeling and traceability system should be conducted at regular intervals[2] to verify that the system:

- Is properly documented and communicated to the appropriate parties;
- Is being implemented by all food workers;
- Meets customer, statutory / regulatory, and Company requirements; and
- Is being monitored and measured for the purpose of continual improvement.

Effectiveness Criteria:

- All food and food ingredients are traceable from their suppliers, throughout Company processes, and on to their destinations.
- Traceability audits are conducted smoothly and quickly, without significant gaps (lags) from one step to the next.
- In the event of a product recall, products are traced promptly and completely and are accounted for in their entirety.

Additional Resources:

A. Food Labeling Guide, A, US Food & Drug Administration, Center for Food Safety and Applied Nutrition (CFSAN), Jun., 1999, revision – http://vm.cfsan.fda.gov/~dms/flg-toc.html. Also see http://www.cfsan.fda.gov/label.html for updates and recent announcements.

B. "FDA 306 Bioterrorism Checklist for Food Related Manufacturers", Operations Technologies, Greenville, SC, USA – http://www.operationstechnologies.com/downloads.htm.

C. Jorgenson, Bill, et al, FDA's Traceback Proposals Offer Challenges and Opportunities, Food Traceability Report, vol. 3, issue 6 (6/1/2003) – http://www.foodtraceabilityreport.com.

D. EU-25 / Food and Agricultural Import Regulations and Standards – EU Traceability Guidelines 2005, USDA Foreign Agricultural Service GAIN Report, 1/21/2005.

E. Traceability in the Food Chain: A Preliminary Study, Food Standards Agency / Food Chain Strategy Division, March, 2002.

---

[1] Once a quarter, at a minimum, is recommended.

[2] ISO guidelines suggest auditing a given function/area at least once every three years.

References:

A. Food Safety Manual

- FSM 7.3.3 – Product Characteristics
- FSM 7.8 – Traceability System

B. Statutory / Regulatory Requirements

Every country and state has (and some larger municipalities have) regulations regarding food labeling. In the *USA*, 21 CFR 101 (Food Labeling) describes food labeling requirements. In *Canada*, the Consumer Packaging and Labeling Act (R.S. 1985, c. C-38) governs food labeling. *Japan's* food-labeling laws include the Law Concerning Prevention of Unfair Gift and Unfair Labeling, which prohibits false or exaggerated labeling, and the Nutrition Improvement Law, which establishes requirements regarding the labeling of products containing nutrients or energy.

The European Commission (EC) issues *directives* to harmonize the laws of the European Union (EU) member states on particular issues – individual member states must enact laws that embody the principles of these directives. Food labeling is enforced by the member state in which the food is sold. Therefore, food manufacturers should consult the specific laws in each member state where the food product is sold.

The Company is required to know and observe all food labeling statutes and regulations pertaining to its end products wherever they are sold.

Due to recent scares like "mad cow disease" (BST) and E. coli O157:H7, laws in most (if not all) countries require that *all* points along the food supply chain show they can *trace* their products to all immediate suppliers and customers (commonly referred to as a "one-up / one-back" requirement).

Food companies doing business in the European Union, for example, must conform to *Article 18 of EC Regulation 178/2002*, which states, in short, that traceability of food and food or feed ingredients must be established at all levels of production, processing, and distribution within a given company. Many countries have such a requirement in place and those that don't are moving in that direction. The *US Bioterrorism Act of 2002, Regulation 306*, is another example of legislation requiring product traceability. Most EU countries also require traceability of genetically modified (GM) foods.

Note that many countries' laws encourage and/or promote *computer-based* recordkeeping, so long as *traceability* of foods and ingredients is maintained.

C. Food Safety Procedure(s)

- FS1010 – FOOD SAFETY RECORDS
- FS1080 – HACCP PLAN MANAGEMENT
- FS1110 – RECEIVING AND INSPECTION
- FS1140 – CONTROL OF MONITORING AND MEASURING

- FS1190 – PRODUCT RECALL

Records:

- FS1130-1 – LOT IDENTIFICATION/PRODUCT TRACEABILITY LOG (EXAMPLE)

  NOTE: The Company is required to have a processing log for every one of its operations (handling, storage, packing, mixing, cooking, cooling, freezing, inspection, sanitation, etc.).
- Identification labels or tags for all products and ingredients
- Receiving logs
- Shipping records

Revision History:

Revision	Date	Description of changes	Requested By
0.0	2/1/2006	Initial Release	

FS1130-1 – LOT IDENTIFICATION / PRODUCT TRACEABILITY LOG (EXAMPLE)

Production Date/Time	Product			Production Line	Ingredients		
	Name/SKU	Quantity	ID/Lot Code		Name/SKU	Quantity	ID/Lot Code

[This page intentionally left blank.]

Doc #: **FS1140**	Title: **CONTROL OF MONITORING AND MEASURING**	Print Date: **2/1/2006**
Rev #: **0.0**	Prepared By:	Date Prepared: **2/1/2006**
Eff. Date: **2/1/2006**	Reviewed By:	Date Reviewed: **2/1/2006**
	Approved By:	Date Approved: **2/1/2006**
Standards: **ISO 22000:2005, clause 8.3**		

Policy: The Company shall monitor and measure foods and food ingredients at all critical control points in its processes to help ensure food safety.

Purpose: This procedure establishes a standard method for the use, calibration, record keeping and maintenance of monitoring and measuring methods and equipment.

Scope: This procedure applies to personnel performing monitoring and measurement activities and the monitoring and measuring devices needed to provide evidence of conformity of product to determined requirements.

Responsibilities:

The Quality Assurance (QA) Manager is responsible for the measuring instrument (device) calibration program.

Purchasing and Quality Assurance are responsible for assuring subcontractor's monitoring and measuring system is adequate to assure food safety requirements.

All personnel are responsible for verifying that instruments they use are within their calibration periods and ensuring that instruments they use are capable of measuring to the required accuracy and tolerance.

Definitions: Calibration – Comparison of a measurement standard or instrument of known accuracy with another standard or instrument to detect, correlate, report, or eliminate by adjustment any variation in the accuracy of the item being compared.

Calibration period – Period during which a certified calibration is valid. For example, if a certifying organization checks and certifies a piece of measuring equipment on 6 March 2006 and the certification expires on 6 September 2006, the calibration period for that instrument is six (6) months.

Critical control point (CCP) – A step at which control can be applied and which step is essential to the prevention or elimination of a food safety hazard or reduction of that hazard to an acceptable level.

Critical limit – Criterion that separates the acceptable from the unacceptable. Critical limits are established to determine whether a CCP remains in control. If a critical limit is exceeded or violated, the affected product is deemed potentially unsafe.

Monitoring – The regular measurement or observation of a critical control point to make sure the product does not go outside of its critical limits.

Monitoring and Measuring Equipment – Devices used to collect data and measure, gauge, test, inspect, or otherwise examine items to determine their compliance with specifications.

Reference Standard – A standard of the highest order of accuracy in a calibration system, establishing the basic accuracy values for that system.

Working Standard – Designated measuring equipment used in a calibration system as a medium for transferring the basic value of reference standards to lower echelon transfer standards or other measuring and test equipment.

Traceability – The ability to relate individual measurement results to national standards or nationally accepted measurement systems through an unbroken chain of comparisons.

Procedure:

1.0 MONITORING AND MEASURING – GENERAL REQUIREMENTS

1.1 All personnel performing monitoring and measurement activities that are needed to provide evidence of conformity of product to food safety requirements must use calibrated equipment. This includes:

- Measuring and testing product;
- Measuring and monitoring processes to demonstrate the process remains within validated parameters; and
- Measuring and monitoring product and processes during development to establish specifications and process parameters.

1.2 All monitoring/measurement personnel shall verify measuring instruments that they use are within their calibration period prior to use.

1.3 A calibration sticker listing the calibration date, the date due (or expiration date), the instrument identification number, and the initials of the person performing the calibration must be attached to each calibrated instrument or device. If it is not practical to attach the sticker directly to the device, it may be attached to the device's container.

1.4 All personnel shall ensure that the monitoring and measuring equipment selected for a particular measure is capable of measuring to the accuracy and tolerance required. In general terms, the measuring device should be four times as accurate as the allowable tolerance of the measurement. The Quality Assurance Manager should be consulted if there are any questions.

2.0 STORAGE, HANDLING, AND MAINTENANCE

2.1 Inspection, measuring and test equipment and measurement standards shall be calibrated and utilized in an environment controlled to the extent necessary to assure continued measurements of required accuracy, giving due consideration to temperature, humidity, vibration, cleanliness, and other controllable factors affecting precision measurement. Measuring instruments shall be used in the conditions for which they were designed by the manufacturer.

2.2 All inspection, measuring and test equipment shall be handled, stored, and transported in a manner that does not adversely affect the calibration or condition of the equipment.

2.3 Each portable measuring tool or instrument shall be kept clean and maintained in a protective container when not in use..

3.0 CALIBRATION SYSTEM

3.1 The Quality Assurance (QA) Department is responsible for maintaining the gage and equipment calibration database. The purpose of this database is to identify equipment location, calibration state, calibration frequency, range of accuracy, and calibration history. The calibration database also identifies those instruments that may be used as secondary standards to calibrate other instruments.

3.2 In the first week of each month, the QA Department will issue notices for Monitoring and Measuring Equipment due for calibration in that month. It is the tool holder's responsibility to return the tool to QA for calibration.

3.3 All gages and equipment are tagged with an identification number, calibration status, and calibration due date. Gages not requiring calibration will be tagged with a "No Calibration Required"

3.4 Gages without tags are to be considered out of calibration and should be returned to QA to be recalibrated.

It is the monitoring and measuring tool holder's responsibility to verify that the calibration sticker is present and that the dates are valid before using the instrument.

3.5 Gage and equipment preventive maintenance is performed in accordance with manufacturer's specifications during calibration. The QA Department will ensure that environmental conditions are suitable for the calibrations, inspections, measurements, and tests being carried out.

3.6 The frequency of calibration for each piece of monitoring and measuring equipment is established by the QA Department and is based on the type of measurements made, the frequency of use, manufacturer specifications, and the calibration history of the device. A default frequency of once every six months is used unless otherwise indicated by Quality. This frequency may be adjusted at any time by the QA Department.

3.7 The QA Department shall document calibrations and maintain records pertaining to calibration, repeatability, and reproducibility of data on FS1140-1 –

CALIBRATION RECORD. Each FS1140-1 shall include, at a minimum, a unique equipment identifier (e.g., equipment name and ID), calibration date, method, standards used, condition of equipment as received, any adjustments or repairs required, calibration measurement data, and identification of the person performing the calibration.

The QA Department shall retain calibration records for the period required by law. Calibration records shall be maintained in a secure file or database (use FS1140-2 – CALIBRATION DATABASE as a guide).

3.8 It is the responsibility of everyone using the measuring and test equipment to alert the QA Department in the event of accidental damage or any irregularities between calibrations.

3.9 Where possible, monitoring and measuring equipment is protected from any adjustments which would invalidate the calibration setting by placing the calibration sticker or a dot of tamper paint where disturbance of the sticker or paint would indicate tampering.

3.10 Calibration may be performed in-house or contracted out to a third party that has been evaluated to ensure that it complies with or exceeds the requirements set forth in ANSI/NCSL Z540 (see Reference A).

3.11 Standards used to calibrate equipment must be traceable to the National Institute of Standards and Technology (NIST) or other recognized national or international standards. If no such standards exist, the basis used for calibration of verification shall be recorded.

3.12 Specific written calibration instructions are used for each piece of equipment or family of equipment. Procedure shall include the following requirements as appropriate for the device:

- The calibration shall be done by comparison to working standards traceable to NIST with an accuracy of 4 to 10 times greater than that of the measuring tool being calibrated.
- The comparisons shall be made at several points across the complete range of use of the device, to ensure linearity.

3.13 If in-house standards are used by a contractor to calibrate device-related measuring equipment, these standards must be documented, used, and maintained the same as other standards.

4.0 OUT-OF-TOLERANCE CONDITIONS

If a gage or instrument is found to be out of tolerance, the QA lab shall immediately pass the out-of-tolerance information to the appropriate supervisor in the department where the instrument is used. The Department and QA management will analyze the effects of the invalid measurements and take immediate, appropriate remedial action for affected in-process or finished products, in accordance with FS1170 – CORRECTIVE ACTION.

5.0 CONTROL OF SUBCONTRACTOR CALIBRATION

Purchasing and Quality Assurance will be responsible for assuring that applicable subcontractor's calibration system conforms to the applicable product plan requirements and to the degree necessary to assure compliance with contractual requirements as indicated in FS1100 – SUPPLIER EVALUATION.

6.0 TEST SOFTWARE

Where test software is used as a form of inspection or testing, it will be checked to prove it is capable of performing the required verification of product acceptability. It will be rechecked at prescribed intervals. Records of the initial check and scheduled rechecks will be maintained in FS1140-1 – CALIBRATION RECORD.

7.0 CALIBRATION PROCESS REVIEW

7.1 QA shall periodically review FS1140-1 – CALIBRATION RECORDS, FS1140-2 – CALIBRATION DATABASE, and other information pertinent to the calibration process and report its findings and observations to Food Safety Management for review. Such reviews shall occur annually, at a minimum.

7.2 Food Safety and QA shall review the process to determine if is being properly implemented and continues to meet requirements. If it is found that the process is not being followed or is inadequate, QA shall revise the process in accordance with FS1170 – CORRECTIVE ACTION.

Effectiveness Criteria:

- Availability of calibrated monitoring and measuring equipment.
- Use of appropriate devices for the measurements required.
- Certification of calibration(s); calibration records.

Additional Resources:

- American National Standards Institute (ANSI) – http://www.ansi.org.
- National Institute of Standards and Technology (NIST), USA – http://www.nist.org.
- Institute for National Measurement Standards (INMS), Canada – http://inms-ienm.nrc-cnrc.gc.ca/en/main_e.php.
- ASTM International – http://www.astm.org.

References:

A. ANSI/NCSL Z540, American National Standard for Calibration – http://www.ncslinternational.org.

B. Food Safety Manual:

- FSM 8.3 - Control of Monitoring and Measuring

C. Food Safety Procedures:
    - FS1100 – SUPPLIER EVALUATION
    - FS1010 – FOOD SAFETY RECORDS
    - FS1170 – CORRECTIVE ACTION

D. The Company's HACCP Plan – Monitoring Procedures (4th principle)

Records:

- FS1140-1 – CALIBRATION RECORD
- FS1140-2 – CALIBRATION DATABASE
- Calibration Laboratory Calibration Certificates
- In-house Calibration Records

Revision History:

Revision	Date	Description of changes	Requested By
0.0	2/1/2006	Initial Release	

FS1140-1 – CALIBRATION RECORD

Equipment Type	Manufacturer
Model	ID No.
Location	
Condition as received	
Accuracy specification	Acceptance range ±
Calibration Standards Used	

Feature / Standard				

Comments

Calibrated By:	Date:

[This page intentionally left blank]

FS1140-2 – CALIBRATION DATABASE

Equipment ID	Equipment Description	Eqpt Mfr	Purchase Date	Calibration Period	Last Calibration Date	Calibration Method	Standard Used	Eqpt Condition as Received	Adjustments-Repairs Required	Measure-ment Data	Person Calibrating

[This page intentionally left blank.]

Doc #: **FS1150**	Title: **CONTROL OF POTENTIALLY UNSAFE FOOD PRODUCT**	Print Date: **2/1/2006**
Rev #: **0.0**	Prepared By:	Date Prepared: **2/1/2006**
Eff. Date: **2/1/2006**	Reviewed By:	Date Reviewed: **2/1/2006**
	Approved By:	Date Approved: **2/1/2006**
Standards: **ISO 22000:2005, clause 7.10.1, 7.10.3**		

Policy: The Company shall ensure that when critical limits for CCP's are exceeded or there is a loss of control of operational PRP's, end products are identified and controlled with regard to their use and release.

Purpose: To outline the procedures for the handling and disposition of nonconforming product. Any nonconforming product item is to be identified, segregated, and documented to prevent inadvertent use. All nonconforming items shall be reviewed to determine the appropriate disposition of each.

Scope: This procedure applies to all nonconforming items found during inspection of received items and materials, during the manufacturing process, and in finished products and products returned by customers.

Responsibilities:

The Food Safety Team Leader is responsible for confirming the nonconformity and determining the disposition of nonconforming material.

The QA Manager is responsible for identifying possible corrective action in the case of nonconforming food items and for ensuring that corrective action is taken, when warranted.

All personnel are responsible for identifying nonconforming – or suspected nonconforming – product or materials and immediately notifying management of the nonconformance.

Definitions: Acceptable level – Level of a safety hazard considered to present an acceptable risk to the consumer. The acceptable level of the hazard in the end product, sometimes referred to as the "target level", should be stated in the product description and set at or below statutory/regulatory limits.

An acceptable level for a hazard at an intermediate step in the product flow diagram may be set higher than that of the final product, provided that the acceptable level in the end product is achieved.

Potentially Unsafe Product – An ingredient, additive, or end product that does not meet food safety specifications. Ingredients or additives that are discovered nonconforming in a production operation and can be brought into a conforming condition as a normal part of the same operation are not considered part of this process.

Critical Control Point (CCP) – Step at which control can be applied and is essential to prevent or eliminate a food safety hazard or reduce it to an acceptable level.

Operational PRP – Operational Prerequisite Program; prerequisite program identified during a hazard analysis as essential to controlling the likelihood of introducing food safety hazards to and/or the contamination or proliferation of food safety hazards in the product or processing environment. (See FS1070 – HAZARD ANALYSIS.)

Recall – Remove a food product from the market because it may cause health problems or possible death. Food manufacturers or distributors typically issue recalls, which may be based on internal or external findings (e.g., consumer complaints).

Withdraw – Same as recall.

Procedure:

1.0 GENERAL

All Company personnel are responsible for ensuring that potentially unsafe products are prevented from entering the food chain unless:

- The food safety hazard of concern can be shown to have been reduced to the defined acceptable levels;
- The food safety hazard of concern is reduced to identified acceptable levels prior to the Company's end product entering the food chain; or
- The product meets the defined acceptable levels of the food safety hazard of concern despite the nonconformity.

2.0 IDENTIFICATION AND SEGREGATION OF NONCONFORMING PRODUCT

2.1 The person identifying the nonconforming product or materials shall clearly and prominently identify the nonconformity by attaching a "Reject" tag to the item or to the bin or container holding the item or items.

2.2 The following information, at a minimum, shall be completed on the Reject tag:

- Material name;
- Identification number (if known);
- Quantity rejected;
- Date and time material rejected;
- Brief description of problem;

- Return Goods Authorization (RGA) number, work order number, or Purchase Order number as applicable; and
- Signature of person rejecting the material.

2.3 Following tagging, the identifier shall segregate the material by placing it in a designated Reject Location or, if this is not possible, clearly separate it from good product and ensure it is clearly identified as Reject material to prevent it from being accidentally used or shipped as acceptable product.

3.0 NONCONFORMANCE REPORT

3.1 The person identifying the nonconformance shall also complete the top section of the FS1150-1 – NONCONFORMANCE REPORT. The following items must be completed, at a minimum:

- Date and time;
- Material or part number;
- Material or part name or description;
- Quantity;
- Problem description;
- Location of product or material;
- Printed name and signature of person identifying the nonconforming material; and
- Work Order or PO number.

3.2 The employee identifying the nonconforming product shall forward the original FS1150-1 – NONCONFORMANCE REPORT to the Food Safety Team Leader for confirmation of the nonconformity and proper disposition of the item.

3.3 The Food Safety Team Leader, upon reviewing the FS1150-1, shall forward a copy to the QA Manager for possible corrective action (see FS1170 – CORRECTIVE ACTION).

3.4 The Food Safety Team Leader and the QA Manager shall determine if corrective action is required to prevent recurrence of the nonconforming condition, in accordance with FS1170 – CORRECTIVE ACTION.

4.0 EVALUATION FOR RELEASE

The food Safety Team Leader will ensure that product affected by the nonconformity will only be released under the following conditions:

- Evidence other than the monitoring system demonstrates that the control measures have been effective
- Evidence shows that the combined effect of the control measures for that particular product complies with the performance intended

- The results of sampling, analysis and/or other verification activities demonstrate that the affected lot of product complies with the identified acceptable levels for the food safety hazard of concern.

5.0 DISPOSITION OF POTENTIALLY UNSAFE PRODUCTS

5.1 The Production Manager, Food Safety Team Leader, and Quality Assurance shall review the evaluation of the nonconforming items. If the affected product cannot be proved safe it will be disposed of on one of the following two ways:

- Reprocessing or further processing within or outside the organization to ensure that the food safety hazard is eliminated or reduced to acceptable levels
- Destruction or disposal as waste in accordance with relevant environmental requirements.

5.2 The Food Safety Team Leader, on determining the disposition of the unsafe product, shall sign the FS1150-1 – NONCONFORMANCE REPORT in the appropriate area.

6.0 RETURNED GOODS

6.1 If the nonconformance applies to product returned from a customer, locate the FS1150-2 – RETURN GOODS AUTHORIZATION FORM and forward the returned product with the FS1150-2 to the QA Manager for analysis.

6.2 A FS1150-1 – NONCONFORMANCE REPORT is not required if an FS1150-2 already exists for the nonconforming item.

7.0 CONTROL PROCESS REVIEW

7.1 The Food Safety Team shall periodically review logs and other records related to control of potentially unsafe product and determine if the control process is being properly implemented and continues to meet Company food safety requirements.

7.2 If it is determined that the process is not meeting such requirements, the Food Safety Team shall make the necessary revisions to the process, in accordance with FS1170 – CORRECTIVE ACTION.

7.3 A third-party audit of the control process shall be conducted at regular intervals, to verify that the process is properly documented and communicated to the appropriate parties, that the process is being implemented consistently, that the process meets the necessary requirements, and that the process is being monitored and measured.

Effectiveness Criteria:

- Prevention of potentially unsafe products reaching customers.
- Adequate records for improvement data analysis.

Additional Resources:

A. None.

References:

A. Food Safety Manual section:
   - FSM 7.9 – Control of Nonconformity

B. Food Safety Procedures:
   - FS1170 – CORRECTIVE ACTION
   - FS1190 – PRODUCT RECALL

C. Statutory/Regulatory Requirements

   Most, if not all, governments have laws in place requiring the proper handling and disposition of food products, ingredients, additives, etc., which are determined to have the potential for harming the consumer. The Company must maintain an awareness of – and observe – the laws governing handling and disposition of unsafe food product in *every* location in which it does business.

Records:

- FS1150-1 – NONCONFORMANCE REPORT
- FS1150-2 – RETURN GOODS AUTHORIZATION FORM
- Nonconforming Material Rework Instruction
- Reject tag

Revision History:

Revision	Date	Description of Changes	Requested By
0.0	2/1/2006	Initial Release	

[This page intentionally left blank]

FS1150-1 – NONCONFORMANCE REPORT

(Attach additional sheets, if necessary.)

Part/Material Number		**NCR #:**
Part/Material Description		**Quantity**
RGA# WO# or PO#	**Location**	**Vendor:**
Nonconformity		
Originator:		**Date:**
Disposition		
❒ **Reprocess** ❒ **Re-grade** ❒ **Hold** ❒ **Destroy/RTV**		
Reprocess Instructions:		
Dispositioned By:		**Date:**
Action Taken:		
By		**Date:**
Corrective Action		
Corrective Action Required?		CAR #
Authorized By:		**Date:**

[This page intentionally left blank]

FS1150-2 – RETURNED GOODS AUTHORIZATION

Date:______________________________ RGA Number:______________________

Customer: ________________________ Reason for Return:__________________

Address: ____

____

____

____

Contact: ____________________ Telephone: ____________________

Part/ Material #	Description	Qty. Shipped	Qty. Returned	Inspected Pass/Fail
________	________________________	____	____	________
________	________________________	____	____	________
________	________________________	____	____	________
________	________________________	____	____	________
________	________________________	____	____	________

Comments:____

____

____

____

____

Received by:__________________________ Date:______________

Accounting Use Only

Credit Memo to be issued: ❒ Yes ❒ No $____________

Amount to be refunded to Customer $____________

Remarks:____

____

____

Approvals

____________________	____________	____________________	____________
Sales/Service Manager	Date	Credit Manager	Date

[This page intentionally left blank]

Document ID: **FS1160**	Title: **INTERNAL AUDIT AND SYSTEM VALIDATION**	Print Date: **2/1/2006**
Revision #: **0.0**	Prepared By:	Date Prepared: **2/1/2006**
Effective Date: **2/1/2006**	Reviewed By:	Date Reviewed: **2/1/2006**
	Approved By:	Date Approved: **2/1/2006**
Standards: **ISO 22000:2005, clauses 7.8 and 8.4**		

Policy: The Company shall conduct regular internal audits of its Food Safety Management System (FSMS).

Purpose: To describe the internal audit process and corrective action as a vital component of the Company's FSMS.

Scope: All Company operations affecting food safety.

Responsibilities:

The Food Safety Manager (or Food Safety Team Leader) is responsible for the internal food safety audit program, selecting personnel to perform internal audits, reviewing the results of any food safety audit, and ensuring that corrective actions are understood and supervising the response to corrective actions.

Food safety audit teams are responsible for conducting internal food safety audits, recommending corrective actions to reduce or eliminate any found nonconformity, and verifying the effectiveness of auditees' responses.

Internal auditors are responsible for conducting complete, detailed, and objective internal audits and reporting their findings to the Food Safety Manager and Top Management.

All personnel are responsible for cooperating with internal food safety auditors during the audit process and for taking appropriate corrective action in response to any deficiency found during the course of the audit.

Definitions: Audit Checklist – A list of questions designed to verify that audit interviewees understand the FSMS and that the FSMS is being effectively implemented.

Audit Plan – Detailed outline of the purpose, scope, objectives and activities for an audit.

Audit Program – Annual summary of areas to be audited, the schedule for each audit, and assigned auditors.

Food Safety Management System (FSMS) – An ordered, well-documented system, the primary goal of which is safe food products for the Company's customer(s).

Procedure:

1.0 INTERNAL AUDIT PROGRAM

1.1 The Food Safety Manager (or Food Safety Team Leader) shall prepare an annual food safety audit program, using FS1160-1 – AUDIT PROGRAM as a guide. The purpose of the audit program is to verify that:

- Activities having an effect on the safety of the Company's product(s) are in compliance with the Company's Food Safety Management System (FSMS);
- The Company's FSMS complies with the most recent version of the ISO 22000 standard; and
- The FSMS is effective in achieving continual improvement of the Company's products/services.

The Food Safety Manager should have an understanding of audit principles, auditor competence, and the application of auditing techniques, in addition to having knowledge of and experience in the Company's processes and food safety concepts and techniques.

1.2 The audit program will specify the areas to be audited and the schedule for each audit. Each area/process having an effect on food safety (e.g., receiving, manufacturing, packaging) shall be audited at least annually.

- Areas/processes may be audited more frequently, based on such factors as past nonconformities, food safety risk consequences, and customer feedback.
- Special audits should be initiated by the Food Safety Manager when a new process is introduced, when equipment or procedures are changed, when a baseline for process improvement is needed, or to assess the effectiveness of any component of the FSMS.

1.3 The Food Safety Manager shall select an audit team for each audit. The size of the audit team may vary according to the area/process being audited. A leader shall be designated for each audit team and leadership assignments should be rotated to provide lead auditor training and experience.

Audit team members may audit any area except those which are included in their primary responsibilities (e.g., an employee whose primary responsibility is "packaging machine operator" must not be allowed to participate in any audit of the Packaging area).

1.4 To be considered competent, internal auditors must be trained in auditing practices and the requirements of the ISO 22000 standard. Before conducting audits on their own, new auditors should perform two audits under an experienced auditor. Audit team leaders must have taken part in at least two audits as an auditor before they may be assigned a lead auditor position.

1.5 The Food Safety Manager shall submit the FS1160-1 – AUDIT PROGRAM to Top Management for review and approval.

- Upon approval of the Audit Program, the Food Safety Manager shall ensure communication of the program to the appropriate parties.

2.0 INTERNAL AUDIT PLANNING

2.1 The assigned audit team leader (Lead Auditor) should prepare the audit plan – using FS1160-2 – AUDIT PLAN as a guide – at least one month prior to the scheduled audit; the lead time required will vary according to the area being audited and the scope of the audit.

- The audit plan should include the purpose, scope, criteria, and objectives for the audit, as well as identify audit date(s) and audit team members. (Unless otherwise specified, the criteria will be the requirements of ISO 22000:2005.)
- A schedule (agenda) for interviewing department personnel should be provided. (When constructing the audit schedule, the audit team leader should be sure to allow time for auditors to review and clarify notes and meet with other team members for discussion).
- Any documents required for the audit shall be identified (e.g., purchase orders from January 1 – March 31 of the current year).
- Any operations or maintenance areas to be visited and observed shall be identified.
- In preparing the plan, the audit team leader should review previous audit reports, corrective action requests (CAR), production histories, complaint files, satisfaction surveys, and any other documents deemed relevant to the audit.

2.2 The audit team leader should prepare checklists for auditors to use in interviewing and observing, using FS1160-3 – FOOD SAFETY AUDIT CHECKLIST EXAMPLE as a guide.

- Audit checklists should be designed to uniquely reflect the activities and critical control points in the area to be audited.

2.3 The Food Safety Manager should review each audit checklist and add questions he/she believes are necessary to adequately evaluate the implementation of the FSMS in the area being audited.

2.4 The Food Safety Manager shall provide the manager of the area being audited sufficient advance notice of the audit to ensure availability of interviewees.

- "Sufficient notice" may vary according to the scope of the audit and the area being audited but should not be less than two weeks, in any case.
- Notification should include oral communication and interoffice mail (e-mail).
- If there are conflicts with the schedule of interviewees, the area manager and Food Safety Manager will resolve those conflicts as appropriate.

3.0 CONDUCTING THE INTERNAL AUDIT

3.1 The audit begins with an opening meeting, as indicated in the audit plan. The opening meeting should be attended by all personnel identified in the plan (other department personnel may attend the meeting, at the discretion of the area manager).

- During the opening meeting, the Lead Auditor (audit team leader) shall introduce the audit team and explain the FS1160-1 – AUDIT PLAN. Copies of the audit plan are distributed (if they have not been previously) and the schedule is confirmed. Questions from department personnel will be answered to the degree the team is able to do so.
- If there are any special (e.g., safety) concerns with the area to be audited, it is the responsibility of the area manager to review any special requirements with the audit team.

3.2 The audit team shall conduct the audit according to the planned schedule to the extent practicable, ensuring a thorough audit. The audit will consist of interviews, document reviews, and observations of area activities.

Auditors should keep in mind that the purpose of the audit is to evaluate objective evidence that the FSMS:

- Complies with the requirements of ISO 22000:2005;
- Is being implemented according to documented procedures and processes;
- Is effective in preventing, eliminating, or controlling food safety risks; and
- Is enabling continual improvement of the Company's products and processes.

3.3 Each member of the audit team shall keep detailed notes of all interviews, observations, and document reviews. Issues beyond the scope of the audit should be noted for follow-up.

3.4 When conducting interviews, audit team members should use the FS1160-3 – FOOD SAFETY AUDIT CHECKLIST EXAMPLE to maintain focus and ensure completeness. There should be enough questions to thoroughly evaluate activities in the audited area, but the team leader should recognize that auditors may not have time to ask all questions.

- Auditors should be flexible. When they find discrepancies, they should ask follow-up questions to clarify their understanding of the situation.

3.5 When the evidence presented is inadequate to demonstrate compliance with a requirement or indicates noncompliance, the auditor shall determine if the issue is a minor or major nonconformity or merely an observation of a potential nonconformity (i.e., it is not a nonconformity but could be, under the right conditions).

- If the evidence points to an observation, the auditor shall discuss this in an audit team review.
- If the evidence points to a nonconformity, the auditor shall fill out an FS1170-2 – NONCONFORMITY REPORT (in accordance with FS1170 – CORRECTIVE ACTION) and present this report in an audit team review.
- The nonconformity must be reviewed by the audit team and a consensus reached, so that the nature and level of the nonconformity is communicated adequately in the team's audit report. The audit team leader shall discuss the nonconformity with the affected manager and obtain his/her concurrence,

indicated on the FS1170-2 by his/her signature. Nonconformity reports will form the basis for each audit report.

4.0 INTERNAL AUDIT REPORTING

4.1 At the conclusion of the auditing activity and prior to any closing meeting, the audit team should meet to integrate the findings, observations, general trends, and specific follow-up issues. During this meeting, the team comes to a conclusion on the degree to which the system audited meets the criteria for the audit.

- Information is collated to collect the strengths, weaknesses, and observations of the entire audit team. Nonconformity reports are reviewed. If any are unsigned, the responsible auditor or audit team leader discusses each with the responsible manager and obtains a signature if possible.
- Possible areas for disagreement are discussed and potential resolutions developed. A follow-up strategy is developed and auditors are assigned responsibility for follow-up.
- If time permits, the Audit Report is drafted, reviewed, and written for presentation at an audit closing meeting.

4.2 Each audit should be concluded with a closing meeting, attended by the audit team, responsible senior management, department personnel (as appropriate), and the Food Safety Manager. During the closing meeting, the audit team leader presents an oral audit report, summarizing:

- The purpose, scope, criteria, and objectives for the audit;
- The overall audit conclusion;
- Strengths and weaknesses;
- Number of nonconformities and observations;
- Recommendations, if any;
- Follow-up activities; and
- Ownership and commitment.

The closing meeting should provide ample time for questions from interested parties.

4.3 A final audit report shall be prepared (using FS1160-4 – EXAMPLE FINAL AUDIT REPORT as a guide), documenting in detail the findings presented during the closing meeting. In addition to the items covered during that meeting, the report will include each of the nonconformity reports generated during the audit, a brief description of any obstacles encountered, a statement of the confidential nature of the contents, and an appendix containing the audit plan and any checklists used and notes generated during the audit.

4.4 When the final audit report is complete and available, it is distributed as directed by Top Management.

5.0 INTERNAL AUDIT FOLLOW-UP

5.1 At a time agreed to by the audit team leader and the manager of the audited department/area/process, the audit team will conduct a follow-up visit, to verify the actions taken as a result of the audit.

5.2 At the time of the follow-up, the manager of the audited area shall demonstrate that actions required by the audit have been implemented and are effective.

6.0 VALIDATION OF THE INTERNAL AUDIT PROCESS

6.1 The internal audit process shall be validated periodically (annually, at a minimum) through an audit. During this validation, evidence will be gathered and evaluated to determine if the internal audit process has been effectively implemented and is producing the desired results.

6.2 Any nonconformity found in the internal audit process by this validation effort should be considered major, in accordance with FS1170 – CORRECTIVE ACTION, since the internal audit process is the *primary* means of assessing the effectiveness of the FSMS.

Effectiveness Criteria:

- Number, frequency, and degree of nonconformity reports (trend data).
- Average time to correct nonconformities.
- Number, frequency, degree of nonconformities.
- No major nonconformities identified by third-party audit (ISO 22000:2005).

References:

A. ISO/FDIS 22000:2005 – Food Safety Management Systems / Requirements for Any Organization in the Food Chain

B. EN/ISO 19011:2002 – Guidelines for Quality and/or Environmental Management Systems Auditing

C. Food Safety Manual

- FSM 8.4 – Food Safety Management System Verification

D. Food Safety Procedure:

- FS1170 – CORRECTIVE ACTION

Records:

- FS1160-1 – AUDIT PROGRAM
- FS1160-2 – AUDIT PLAN
- FS1160-3 – FOOD SAFETY AUDIT CHECKLIST EXAMPLE
- FS1160-4 – FINAL AUDIT REPORT
- FS1170-1 – CORRECTIVE ACTION REQUEST
- FS1170-3 – NONCONFORMITY REPORT

- Auditor Register
- Auditor Log

Revision History:

Revision	Date	Description of changes	Requested By
0.0	2/1/2006	Initial Release	

[This page intentionally left blank.]

FS1160-1 – AUDIT PROGRAM

	(year)					
Audit Scope	**Jan-Feb**	**Mar-Apr**	**May-Jun**	**Jul-Aug**	**Sep-Oct**	**Nov-Dec**
• Document Control • Food Safety Records • Management Responsibility	Plan Audit Team (lead auditor, auditors)	Audit	Follow Up			
• Competence, Awareness and Training • Job Descriptions		Plan Audit Team (lead auditor, auditors)	Audit	Follow Up		
• Prerequisite Programs • Hazard Analysis Preparation • Hazard Analysis • HACCP Plan Management			Plan Audit Team (lead auditor, auditors)	Audit	Follow Up	
• Purchasing • Supplier Evaluation • Receiving and Inspection				Plan Audit Team (lead auditor, auditors)	Audit	Follow Up
• Manufacturing • Identification, Labeling, and Traceability • Control of Monitoring and Measuring • Control of Nonconforming Product • Product Recall	Follow Up				Plan Audit Team (lead auditor, auditors)	Audit

	(year)					
Audit Scope	**Jan-Feb**	**Mar-Apr**	**May-Jun**	**Jul-Aug**	**Sep-Oct**	**Nov-Dec**
• Internal Audit and System Validation • Corrections; Corrective Action • Continuous Improvement	Audit	Follow Up				Plan Audit Team (lead auditor, auditors)

FS1160-2 – AUDIT PLAN

Purpose: To evaluate the effectiveness of the HACCP Plan and supporting programs

Audit Scope:
- Prerequisite Programs
- Hazard Analysis Preparation
- Hazard Analysis
- HACCP Plan Management

Audit Team:
- Bill Oakes Lead Auditor
- Tom Jenkins Auditor
- Sandra Okumi Auditor
- Clarence Thompson Auditor

Criteria:
- ISO 22000:2005
- Our Company's Food Safety Management System (FSMS)

Objective: To determine readiness for the annual Food Safety Surveillance audit

Date: July 7, 2006

Area Personnel:

Rob Portman	Manufacturing Manager
Nancy Givens	Food Safety Manager
Brenda Croft	Product Development
Dean Phelps	Food Safety Lab Supervisor
George Mencke	Quality Manager
Ann Thomas	Purchasing Manager
Mohammad Chowdhury	Process Engineer

AGENDA

Time	Activity	Auditor	Area Personnel
08:00 – 08:30	Opening Meeting	Audit Team	All
08:45-09:30	Prerequisite Programs Hazard Analysis Preparation Hazard Analysis Preparation	Oakes, Okumi Jenkins Thompson	Portman, Mencke Givens Croft
09:30-09:45	Break		
09:45-11:00	Manufacturing Tour	Audit Team	Portman, Chowdhury
11:15-12:00	Hazard Analysis Hazard Analysis Hazard Analysis	Okumi, Jenkins Thompson Oakes	Portman, Mencke Thomas Givens, Croft Chowdhury
12:15-13:00	Lunch	Audit Team	
13:00-14:00	HACCP Plan Management HACCP Plan Management	Oakes, Thompson Jenkins, Okumi	Givens, Portman Mencke, Phelps
14:15-15:30	Audit Report Preparation	Audit Team	
15:30-16:00	Closing Meeting	Audit Team	All

This page intentionally left blank

FS1160-3 – FOOD SAFETY AUDIT CHECKLIST EXAMPLE

AREA AUDITED: ______________________________ DATE: __________

ISO 22000 REQUIREMENT	STATUS	ACTION/COMMENTS
1.0 FOOD SAFETY MANAGEMENT SYSTEM		
1.1 GENERAL REQUIREMENTS		
a) Is the scope of the food safety management system adequately defined?		
b) What evidence exists to demonstrate that food safety hazards are identified, evaluated and controlled to prevent harm to the consumer?		
c) Is appropriate information communicated throughout the food chain regarding safety issues related to products?		
d) Is information concerning development, implementation and updating of the food safety management system communicated throughout the organization to ensure food safety?		
e) Is there evidence that the food safety management system is evaluated and update periodically?		
1.2 DOCUMENTATION REQUIREMENTS		
1.2.1 General		
a) Does a food safety policy and related objectives exist?		
b) Do documented procedures and records exist as required by this standard exist?		
c) Is there evidence of documents needed by the organization to ensure the effective development, implementation, and updating of the food safety management system?		
1.2.2 Document Control		
a) Is there a documented procedure for control of documents?		
b) Is there evidence that documents are reviewed and updated as		

ISO 22000 REQUIREMENT	STATUS	ACTION/COMMENTS
required?		
1.2.3 Control of Records		
a) Is there a documented procedure for control of records?		
b) Are all records legible?		
c) Are records easily retrievable?		
d) Is there evidence that records are properly identified, stored, protected, and disposed of?		
2.0 MANAGEMENT RESPONSIBILITY		
2.1 MANAGEMENT COMMITMENT		
a) Has top management communicated the importance of food safety requirements?		
b) Has a food safety policy been established and communicated?		
c) Has management demonstrated that food safety is supported by the business objectives?		
d) What evidence of management commitment exists (e.g., management reviews, resources)?		
2.2 FOOD SAFETY POLICY		
a) Is the food safety policy appropriate to the role of the organization in the food chain?		
b) Does the Company's food safety policy conform to statutory/ regulatory requirements *and* mutually agreed upon food safety requirements of customers?		
c) Is the food safety policy communicated, implemented, and maintained at all levels of the organization?		
d) What evidence exists to show that the food safety policy is regularly reviewed for continued suitability?		
e) Is there evidence that the Company's food safety policy addresses communication adequately?		
f) Do measurable objectives exist which support the food safety		

ISO 22000 REQUIREMENT	STATUS	ACTION/COMMENTS
policy?		
2.3 FOOD SAFETY MANAGEMENT SYSTEM PLANNING		
a) What evidence exists to show that planning of the food safety management system is carried out to meet the requirements given in 4.1, as well as the Company's food safety objectives?		
b) What controls are in place to assure the integrity of the food safety management system when changes to the food safety management system are planned and implemented?		
2.4 RESPONSIBILITY AND AUTHORITY		
a) Are responsibilities and authorities for food safety management clearly defined and communicated within the Company/department?		
b) Are there designated personnel with defined responsibility and authority to initiate and record actions regarding food safety issues?		
c) What evidence exists to show that all personnel understand their responsibility for reporting problems with the food safety management system to designated personnel?		
2.5 FOOD SAFETY TEAM LEADER		
a) Has a Food Safety Team Leader (Food Safety Manager) been appointed and given responsibility and authority to manage a food safety team and organize its work?		
b) Is the Food Safety Team Leader ensuring that relevant training and education of the food safety team members?		
c) Is there evidence to indicate the Food Safety Team Leader has established, implemented, and is maintaining and updating the		

ISO 22000 REQUIREMENT	STATUS	ACTION/COMMENTS
FSMS?		
d) Is there evidence to indicate the Food Safety Team Leader is reporting directly to Top Management on the effectiveness and suitability of the FSMS?		
2.6 COMMUNICATION		
2.6.1 External Communications		
a) Has the organization established and does it maintain effective communications with suppliers and contractors?		
b) Has the organization established and maintained effective communications with customers or consumers (in particular, in relation to product information, specific storage requirements, shelf life, and customer feedback, including customer complaints)?		
c) Has the organization established and maintained effective communications with statutory and regulatory authorities?		
d) Has the organization established and maintained effective communications with other organizations that have an impact on the effectiveness or updating of the Food Safety Management System?		
e) Is there evidence that external communications are recorded and such records are maintained?		
f) Are food safety requirements from statutory and regulatory authorities and customers available where needed at the workplace?		
g) Have designated personnel been given responsibility and authority to communicate any information concerning food safety externally?		
2.6.2 Internal Communications		
a) What evidence exists to show that the organization has implemented and maintained effective communications with personnel on		

ISO 22000 REQUIREMENT	STATUS	ACTION/COMMENTS
issues having an impact on food safety?		
b) Is there evidence of a process by which the food safety team is informed (and, if applicable, evidence of the process being followed) of changes to the following:		
• Products or new products?		
• Raw materials, ingredients, and services?		
• Production systems and equipment?		
• Production premises, location of equipment, and surrounding environment?		
• Cleaning and sanitation programs?		
• Packaging, storage, and distribution systems?		
• Personnel qualification levels and/or allocation of responsibilities and authorizations?		
• Statutory and regulatory requirements?		
• Knowledge regarding food safety hazards and control measures?		
• Customer, sector, and other requirements that the Company observes?		
• Relevant inquiries from external interested parties?		
• Complaints indicating food safety hazards associated with the product?		
• Other conditions that have an impact on food safety?		
c) Is there evidence that information regarding changes affecting food safety is included in FSMS updates?		
2.7 EMERGENCY PREPAREDNESS AND RESPONSE		
a) Does the Company have in place ***procedures for managing potential emergency situations and accidents*** that may impact food safety and which are relevant to the		

ISO 22000 REQUIREMENT	STATUS	ACTION/COMMENTS
role of the organization in the food chain?		
b) Is there evidence to indicate that when an emergency situation arose:		
• The affected department(s) followed the predetermined emergency procedure?		
• Following the predetermined emergency procedure prevented or minimized hazards, disruptions, etc.?		
2.8 MANAGEMENT REVIEW		
2.8.1 General		
a) Is there evidence that top management reviews the organization's food safety management system at planned intervals to ensure its continuing suitability, adequacy, and effectiveness?		
b) Does this review include assessing opportunities for improvement and the need for change to the food safety management system, including the food safety policy?		
c) Are records of management reviews maintained?		
2.8.2 Review input		
a) Is there evidence to show that input to management reviews includes information on:		
• Follow-up actions from previous management reviews?		
• Analysis of results of verification activities?		
• Changing circumstances that can affect food safety?		
• Emergency situations, accidents, and recalls?		
• Reviewing results of system updating activities?		
• Review of communication activities, including customer feedback?		
• External audits or inspections?		
b) Is input presented in a manner that enables Top Management to relate		

ISO 22000 REQUIREMENT	STATUS	ACTION/COMMENTS
it to the stated objectives of the Food Safety Management System?		
2.8.3 Review output		
Is there evidence that the output from the management review includes decisions and actions related to:		
a) Food safety assurance?		
b) Improving FSMS effectiveness?		
c) Resource needs?		
d) Revisions of the Company's food safety policy and related objectives?		
3.0 RESOURCE MANAGEMENT		
3.1 PROVISION OF RESOURCES		
Is there evidence that the Company is providing adequate resources for the establishment, implementation, maintenance, and updating of its FSMS?		
3.2 HUMAN RESOURCES		
3.2.1 General		
a) What evidence demonstrates that the food safety team and other personnel carrying out activities having an impact on food safety are ***competent*** and have appropriate ***education***, ***training***, ***skills***, and ***experience***?		
b) Where expert assistance is required to develop, implement, operate, or assess the Company's FSMS, are there records of agreement or contracts defining the responsibility and authority of ***external experts***?		
3.2.2 Competence, awareness and training – Has the Company:		
a) Identified the necessary competencies for personnel whose activities have an impact on food safety?		
b) Provided training or taken other action to ensure personnel have the necessary competencies?		
c) Ensured that personnel responsible for monitoring, corrections, and corrective actions of the food safety management system are trained?		

ISO 22000 REQUIREMENT	STATUS	ACTION/COMMENTS
d) Evaluated the implementation and effectiveness of a), b) and c)?		
e) Ensured that all personnel are aware of the relevance and importance of their individual activities in contributing to food safety?		
f) Ensured that the requirement for effective communication is understood by ***all*** personnel whose activities have an impact on food safety?		
g) Maintained appropriate records of training and actions described in b) and c)?		
3.3 INFRASTRUCTURE		
What evidence shows the Company is providing resources for establishing and maintaining the ***infrastructure*** (processing equipment, office equipment, utilities, services, etc.) it needs to implement ISO 22000 requirements?		
3.4 WORK ENVIRONMENT		
Is there evidence that the Company is providing adequate resources for establishment, management, and maintenance of the work environment (e.g., building maintenance plans & schedule) needed to implement ISO 22000 requirements?		
4.0 PLANNING AND REALIZATION OF SAFE PRODUCTS		
4.1 GENERAL		
a) Is there evidence that the organization has planned and developed processes needed for the realization of safe products?		
b) What is the evidence that the organization has implemented, operated, and ensured the effectiveness of the planned activities and any changes to those activities?		

ISO 22000 REQUIREMENT	STATUS	ACTION/COMMENTS
4.2 PREREQUISITE PROGRAMS (PRPs)		
4.2.1 **Has the Company developed, implemented, and maintained prerequisite programs (PRPs)** for the purpose of preventing, eliminating, or reducing:		
a) The likelihood of introducing food safety hazards to the product through the work environment?		
b) Biological, chemical, and physical contamination of Company product(s), including cross-contamination between products or product lines?		
c) Food safety hazard levels in the product and product processing environment?		
4.2.2 Are the Company's PRPs:		
a) Appropriate to the organization's needs regarding food safety?		
b) Appropriate to the size and type of the operation and the nature of the products being manufactured and/or handled by the Company?		
c) Implemented across the entire production system (as programs applicable in general or to specific Company products or operational lines)?		
d) Approved by the Food Safety Team?		
4.2.3 Has the organization considered and utilized appropriate information (e.g., statutory/regulatory requirements, customer requirements, recognized guidelines (Good Practices, standard operating procedures, etc.), Codex Alimentarius Commission principles and codes of practice, national, international or sector standards) **in developing its PRPs?**		
a) Is there evidence to show the Company considered the following when establishing its PRPs:		
• Construction and layout of buildings and associated utilities?		
• Lay-out of premises, including workspace and employee facilities?		
• Supplies of air, water, energy and other utilities?		
• Supporting services, including waste and sewage disposal?		
• The suitability of equipment and its accessibility for cleaning, maintenance and preventative maintenance?		
• Management of purchased materials (e.g., raw material,		

ISO 22000 REQUIREMENT	STATUS	ACTION/COMMENTS
ingredients, chemicals, and packaging), supplies, waste disposal, and handling of products?		
• Measures for the prevention of cross-contamination?		
• Cleaning and sanitizing?		
• Pest control?		
• Personal hygiene?		
• Other aspects as appropriate?		
b) What evidence shows that verification of PRPs has been planned?		
c) What evidence shows that PRPs are modified as needed?		
d) Are there documents specifying how activities included in PRPs are to be managed?		
4.3 PRELIMINARY STEPS TO ENABLE HAZARD ANALYSIS		
4.3.1 General		
Are there records indicating that all relevant information needed to conduct the hazard analysis has been collected, maintained, updated, and documented?		
4.3.2 Food Safety Team		
a) Is there documentary evidence of a Food Safety Team and Team Leader being appointed for hazard analyses conducted during the period under review?		
b) Is there evidence to show that Food Safety Teams had the required combination of multidisciplinary knowledge *and* experience in developing and implementing the Company's FSMS?		
4.3.3.1 Raw Materials, Ingredients, and Product-Contact Materials		
a) Have all raw materials, ingredients, and product-contact materials been described in documents to the extent needed to conduct hazard analyses (including the following, as appropriate):		
• Biological, chemical, and physical characteristics of raw materials, ingredients, and product-contact materials?		
• Composition of formulated ingredients, including additives		

ISO 22000 REQUIREMENT	STATUS	ACTION/COMMENTS
and processing aids?		
• Origin (source) of materials?		
• Method(s) of production?		
• Packaging/delivery methods?		
• Appropriate storage conditions and recommended shelf life?		
• Preparation and/or handling before use or processing?		
• Food safety-related acceptance criteria or specifications of purchased materials and ingredients appropriate to their intended uses?		
b) Has the organization identified statutory/regulatory requirements related to the safety of raw materials, ingredients, and product-contact materials?		
c) Are descriptions of raw materials, ingredients, and product-contact materials up-to-date?		
4.3.3.2 End Product Characteristics		
a) Have end product characteristics been documented to the extent needed to conduct all hazard analyses in the period being audited, including – but not limited to – the following:		
• Product name or similar identification?		
• Composition (e.g., ingredients, agents)?		
• Biological, chemical, and physical characteristics relevant to food safety?		
• Intended shelf life and storage conditions?		
• Packaging?		
• Labeling relating to food safety and/or instructions for handling, preparation, and usage?		
• Distribution methods?		
b) Has the organization identified statutory/regulatory food safety requirements related to end product characteristics?		
c) Have end product descriptions been kept up-to-date?		
4.3.4 Intended Use		
a) Have the intended use *and* the reasonably expected handling or		

ISO 22000 REQUIREMENT	STATUS	ACTION/COMMENTS
mishandling of the end product been described in documents to the extent needed to conduct the hazard analysis (see 7.4)?		
b) What evidence indicates that groups of users (consumers) have been identified for each product?		
c) Have consumer groups known to be especially vulnerable to specific food safety hazards (e.g., food allergies) been considered?		
d) Are the descriptions of user (consumer) groups up-to-date?		
4.3.5 Flow Diagrams, Process Steps and Control Measures		
4.3.5.1 Flow Diagrams		
a) Do flow diagrams exist for the products or process categories covered by the Food Safety Management System?		
b) Do the flow diagrams provide a basis for evaluating the possible occurrence, increase, or introduction of food safety hazards?		
c) Are flow diagrams clear, accurate, and sufficiently detailed? Do they include:		
• The sequence and interaction of all steps in the operation?		
• Any outsourced processes and subcontracted work?		
• Where raw materials, ingredients, and intermediate products enter the flow?		
• Where reworking and recycling take place?		
• Where end products, intermediate products, by-products, and waste are released or removed?		
d) What evidence shows that the Food Safety Team has verified the accuracy of flow diagrams by conducting document walkthroughs and on-site verifications?		
4.3.5.2 Description Of Process Steps and Control Measures		
a) Have existing control measures, process parameters, and procedures that may influence food safety been described to the extent needed to conduct a hazard analysis?		

ISO 22000 REQUIREMENT	STATUS	ACTION/COMMENTS
b) Have external requirements (that may impact the choice and the rigorousness of the control measures) been described?		
c) Is there evidence that the descriptions of process steps and control measures have been maintained and/or updated?		
4.4 HAZARD ANALYSIS		
4.4.1 General		
a) What evidence indicates that the Food Safety Team conducted the hazard analysis and determined:		
• Which hazards needed to be controlled?		
• The degree of control required to ensure food safety?		
• What (combination of) control measures is required?		
b) What was the date of the hazard analysis?		
c) Does the Company have guidelines for reviewing/conducting hazard analyses, to ensure hazard analysis information is relevant and up-to-date?		
4.4.2 Hazard identification and determination of acceptable levels		
a) Have all food safety hazards that are reasonably expected to occur been identified and recorded?		
b) What evidence exists to show hazard identification was based on:		
• The preliminary information and data collected according to 7.3?		
• Experience?		
• External information including, to the extent possible, epidemiological and other historical data?		
• Information from the food chain on safety hazards that may be relevant to the safety of the Company's end products, intermediate products, and the food at consumption?		
• Identification of the step(s) (from raw materials,		

ISO 22000 REQUIREMENT	STATUS	ACTION/COMMENTS
processing, and distribution) at which each food safety hazard may be introduced?		
c) Is there evidence to show that when identifying hazards, the Food Safety Team considered:		
• The steps preceding *and following* the specified operation?		
• The process equipment, utilities/services, and surroundings?		
• The preceding *and following* links in the food supply chain?		
d) Has an acceptable level of each identified hazard been determined?		
4.4.3 Hazard assessment		
a) Has a hazard assessment been conducted for each food safety hazard identified, to determine:		
• Whether prevention, elimination, or reduction of the hazard to acceptable levels is *essential* to the production of a safe food?		
• Whether hazard control is needed to enable the defined acceptable levels to be met?		
b) Is there evidence that each food safety hazard has been evaluated for the *possible severity* of adverse health effects and the *likelihood* of its occurrence?		
c) Has the hazard assessment methodology used been described and the results of the food safety hazard assessment recorded?		
4.4.4 Selection and Assessment of Control Measures		
a) Based on the hazard assessment, is there evidence that the selected combination of control measures is capable of preventing, eliminating, or reducing identified food safety hazards to defined acceptable levels?		
b) Has each selected control measure been reviewed with respect to its effectiveness against identified food safety hazard(s)?		
c) Is there evidence that each selected control measure been categorized as to whether it needs to be managed through operational PRP's or by the HACCP plan regarding:		
• Its effect on identified food safety hazards, relative to the		

ISO 22000 REQUIREMENT	STATUS	ACTION/COMMENTS
strictness applied?		
• Its feasibility for monitoring (e.g., ability to be monitored in a timely manner, enabling immediate corrective actions)?		
• Its place within the system relative to other control measures?		
• The likelihood of a control measure failing to function or functioning in a way that leads to significant processing variability?		
• The severity of the consequence(s) if the control measure fails to function?		
• Whether the control measure is specifically established and applied to prevent, eliminate, or reduce hazard(s) to acceptable levels?		
• Synergistic effects (i.e., interaction that occurs between two or more measures, resulting in their combined effect being higher than the sum of their individual effects)?		
d) Have the methodology and parameters used for this categorization been described in documents and the results of the hazard assessment recorded?		
4.5 OPERATIONAL PREREQUISITE PROGRAMS		
Have operational PRPs been documented and do these documents include the following information for each program:		
a) Food safety hazards to be controlled by the program?		
b) Control measures?		
c) Monitoring procedures that demonstrate that operational PRPs are in place?		
d) Corrections and corrective actions to be taken if monitoring shows that operational PRPs are not in control?		
e) Responsibilities and authorities regarding each operational PRP?		
f) Records of monitoring and measurement?		

ISO 22000 REQUIREMENT	STATUS	ACTION/COMMENTS
4.6 ESTABLISHING THE HACCP PLAN		
4.6.1 The HACCP Plan – Is it documented *and* does such documentation include the following information for each identified critical control point (CCP):		
a) Food safety hazards to be controlled at the CCP?		
b) Control measures?		
c) Critical limits?		
d) Monitoring procedures?		
e) Corrections and corrective actions to be taken if critical limits are exceeded?		
f) HACCP plan responsibilities and authorities?		
g) Records of monitoring/ measurement?		
4.6.2 Identification of Critical Control Points (CCPs)		
Have critical control points been identified for *all* control measures identified for *each* hazard?		
4.6.3 Determination of Critical Limits for Critical Control Points		
a) Have critical limits been determined for each CCP?		
b) What evidence shows that established critical limits will ensure that identified acceptable levels of food safety hazards in the Company's end products are not exceeded?		
c) Are the critical limits *measurable*?		
d) Is the *rationale* for identified critical limits documented?		
e) When critical limits are based on *subjective* data (e.g., visual inspection), are they supported by instructions and/or education and training?		
4.6.4 System for Monitoring of Critical Control Points		
a) Is there evidence of a monitoring system for each CCP which demonstrates that the CCP is in control?		
b) Do monitoring system procedures, instructions, and records cover the following:		
• Measurements or observations that provide results within an adequate time frame?		

ISO 22000 REQUIREMENT	STATUS	ACTION/COMMENTS
• Monitoring devices used?		
• Applicable calibration methods (see 8.3)?		
• Monitoring frequency?		
• Responsibilities and authority related to CCP monitoring and evaluation?		
• Record requirements and methods?		
c) What evidence shows that CCP monitoring methods and frequency are capable of determining *when* critical limits have been exceeded *in time* for the product to be isolated or controlled *before* it is used or consumed?		
4.6.5 Actions When Monitoring Results Exceed Critical Limits		
a) Does the HACCP plan specify the planned corrections and corrective actions to be taken when critical limits are exceeded?		
b) What evidence is there that corrective actions specified in the HACCP plan will ensure that the cause of nonconformity is identified, that parameters controlled at the CCP are brought back under control, and that recurrence of the nonconformity will be prevented?		
c) Are there documented procedures for handling potentially unsafe products?		
d) Is there evidence that these procedures ensured that potentially unsafe products were not released until they could be evaluated?		
4.7 UPDATING OF PRELIMINARY INFORMATION AND DOCUMENTS		
What evidence demonstrates that the Company is updating the following, when necessary:		
a) Product characteristics?		
b) Intended use of its products?		
c) Flow diagrams?		
d) Process steps?		
e) Control measures?		
f) HACCP plans?		

ISO 22000 REQUIREMENT	STATUS	ACTION/COMMENTS
g) Procedures and instructions specifying PRPs?		
7.8 VERIFICATION PLANNING		
a) Are the purpose, methods, frequencies, and responsibilities for verification activities defined?		
b) Is there evidence that verification activities have taken place that would confirm:		
• Prerequisite programs have been implemented?		
• Input to hazard analyses is continually updated?		
• Operational PRPs and elements of the HACCP plan are being implemented and are effective?		
• Food safety hazards are within identified acceptable levels?		
• Other procedures required by the organization to ensure food safety are being implemented and are effective?		
c) Are all verification results recorded and communicated to the appropriate Food Safety Team?		
d) If system verification test samples showed *unacceptable levels of a food safety hazard*, is there evidence to show are affected product lots were properly handled as "potentially unsafe"?		
4.9 TRACEABILITY SYSTEM		
a) Is there evidence that the Company established and applied a *traceability system* to enable identification of product lots *and* their relation to batches of raw materials, processing, and delivery records?		
b) Where is the evidence that the traceability system enabled identification of incoming material, from the immediate supplier(s) through the initial distribution of the Company's end product?		
c) Are traceability records maintained for a defined period, for system		

ISO 22000 REQUIREMENT	STATUS	ACTION/COMMENTS
assessment to enable the handling of potentially unsafe products and in the event of product recall?		
4.10 CONTROL OF NONCONFORMITY		
4.10.1 Corrections		
a) Is there evidence that when critical limits for a CCP are exceeded or there is a loss of control of operational PRPs, the affected products are being identified and controlled with regard to use and release?		
b) Is there a documented procedure defining:		
• Identification and assessment of affected end products, to determine their proper handling?		
• A review of the corrections carried out?		
c) Is there evidence that products *manufactured under conditions not conforming to operational PRPs* are being evaluated with respect to the possible food safety consequences?		
d) Are all corrections approved by the responsible person(s) and recorded, together with information on the nature of the nonconformity, its causes, and its consequences, including information needed for traceability of nonconforming lots?		
4.10.2 Corrective actions		
a) Is there evidence to show that data derived from the monitoring of operational PRPs and CCPs are evaluated by designated persons who possess sufficient knowledge and authority to initiate corrective actions?		
b) Is there evidence that corrective actions are initiated when critical limits are exceeded or when there is a lack of conformity with operational PRPs?		
c) Are there documented procedures that specify actions to identify and eliminate the cause of		

ISO 22000 REQUIREMENT	STATUS	ACTION/COMMENTS
nonconformities, prevent recurrence, and bring the process (system) back into control once a nonconformity is encountered?		
d) Do the specified actions in the procedures in c) include:		
• Reviewing nonconformities (including customer complaints)?		
• Reviewing monitoring results for evidence of a trend toward loss of control?		
• Determining the cause(s) of nonconformities?		
• Evaluating the need for action to ensure that nonconformities do not recur?		
• Determining and implementing the actions needed?		
• Recording the results of corrective actions taken?		
• Reviewing corrective actions taken to ensure their effectiveness?		
e) Are there records of *corrections*?		
f) Are there records of *corrective actions*?		
4.10.3 Handling of Potentially Unsafe Products		
4.10.3.1 General		
a) Is there evidence that the Company is handling nonconforming product by taking action to prevent it from entering the food chain?		
b) If the Company did *not* take such action, is there evidence that:		
• The food safety hazard of concern has been reduced to defined acceptable levels?		
• The food safety hazard will be reduced to defined acceptable levels (see 7.4.2) prior to the product entering the food chain?		
• The product meets the defined acceptable levels of the food safety hazard, despite the nonconformity?		
c) Is there evidence that all lots of product that *may have been affected by a nonconformity* were held in the Company's control until they		

ISO 22000 REQUIREMENT	STATUS	ACTION/COMMENTS
were evaluated?		
d) Is there evidence that when a product left the Company's control *and was subsequently determined to be unsafe*, the Company notified relevant interested parties and initiated a *recall*?		
e) Are all controls (and related responses and authorizations) for dealing with potentially unsafe products *documented*?		
7.10.3.2 Evaluation for Release – Is each lot of product affected by a nonconformity *only* released as safe when one or more of the following conditions applies:		
a) Is there evidence *other than the Company's monitoring systems* that control measures have been effective?		
b) Is there evidence that the *combined effect* of the control measures for that particular product complies with the intended performance/ result (i.e., one or more control measures did not work as intended, yet the cumulative effect of the measures that *did* work resulted in a safe product)?		
c) The results of sampling, analysis, and/or other verification activities demonstrate that the affected lot of product complies with the identified acceptable levels for the food safety hazard(s) concerned?		
4.10.3.3 Disposition of Nonconforming Products – Is there evidence that if a product lot was *not acceptable* for release, it was handled by *one* of the following activities:		
a) Reprocessing or further processing within – or outside of – the organization took place, which ensured that the food safety hazard was eliminated or reduced to acceptable levels?		
b) The nonconforming product was destroyed or disposed of as waste?		
4.10.4 Recalls (Withdrawals)		
a) What evidence demonstrates that if end products were identified as unsafe, they were *completely* recovered *in a timely fashion*?		
• Has Top Management appointed personnel with authority to *initiate* recalls and appointed personnel		

ISO 22000 REQUIREMENT	STATUS	ACTION/COMMENTS
responsible for *executing* recalls?		
• Has the organization established and maintained a documented procedure for:		
o Notifying relevant, interested parties (i.e., statutory/regulatory authorities, the Company's customers, and consumers)?		
o Handling recalled products and affected lots of product still in stock?		
o The sequence of actions to be taken in the event of a recall?		
b) What evidence exists to demonstrate that recalled products are secured or held under supervision until their final disposition can be determined?		
c) Are the cause, extent, and results of recalls recorded and reported to Top Management during management reviews (see 5.8.2)?		
d) Is there evidence of the effectiveness of the product recall program?		
5.0 VALIDATION, VERIFICATION, AND IMPROVEMENT OF THE FOOD SAFETY MANAGEMENT SYSTEM		
5.1 GENERAL		
Is there evidence that the processes needed to *validate* control measures and *verify and improve* the FSMS have been *planned and implemented*?		
5.2 VALIDATION OF CONTROL MEASURE COMBINATIONS		
a) Have control measures included in operational PRPs and HACCP plans been validated, assuring that:		
• Selected control measures, separately or in combination, are capable of achieving the intended control of food safety hazards for which they are designated?		
• Control measures are effective and capable, separately or in		

ISO 22000 REQUIREMENT	STATUS	ACTION/COMMENTS
combination, of ensuring control of identified food safety hazards so that the Company's end products meet defined acceptable levels?		
b) If the result of a validation showed that one or both of the above elements *could not be confirmed*, is there evidence that the control measure or combination of measures has been or is being modified and reassessed?		
5.3 CONTROL OF MONITORING AND MEASURING		
a) Is there evidence that the specified monitoring and measuring methods and equipment are adequate to ensure performance of monitoring and measuring procedures?		
b) Where necessary to ensure valid results, is there evidence that measuring equipment and methods used have been:		
• Calibrated or verified at specified intervals or prior to use against measurement standards traceable to international or national measurement standards bodies?		
• Adjusted or readjusted as necessary?		
• Identified to enable the calibration status to be determined?		
• Safeguarded from adjustments that would invalidate the measurement results?		
• Protected, where possible, from damage and deterioration (or if such protection is not possible, that a replacement schedule that prevents damage and deterioration is established and implemented)?		
c) Are there records of calibration and verification results?		
d) Does the Company assess the validity of previous measurement results when the equipment or process is found not to conform to requirements?		

ISO 22000 REQUIREMENT	STATUS	ACTION/COMMENTS
e) If the measuring equipment is found to be nonconforming, does the organization have an established plan to take action appropriate to the equipment and any product affected? Is there evidence that the plan is being implemented?		
f) Are there records of such validity assessments in e) and their resulting actions?		
g) When used in the monitoring and measurement of specified requirements, was the ability of computer software to satisfy the intended application requirement(s) confirmed?		
5.4 FOOD SAFETY MANAGEMENT SYSTEM VERIFICATION		
5.4.1 Internal Audit		
a) Is there evidence that the Company is conducting internal audits at planned intervals to determine whether the FSMS:		
• Conforms to planned arrangements, FSMS requirements established by the Company, and the requirements of ISO 22000?		
• Is effectively implemented and updated as needed?		
b) Is there an audit program which takes into consideration the importance of the processes and areas to be audited, as well as any updating actions resulting from previous audits?		
c) Are internal audit criteria, scope, frequency, and methods defined prior to the audit?		
d) What evidence exists to show that the selection of auditors and the conduct of audits ensure the objectivity and impartiality of the audit process?		
e) What controls exist to assure that auditors do not audit their own work?		
f) Is there a documented procedure for planning and conducting audits, reporting results, and maintaining		

ISO 22000 REQUIREMENT	STATUS	ACTION/COMMENTS
audit records?		
g) Has management responsible for the area(s) being audited taken actions *without undue delay* to eliminate detected nonconformities and their causes?		
h) Is there evidence that audit teams follow up their audits within the prescribed period of time?		
i) Did follow-up activities include verification of any corrective actions taken and the reporting of the verification results?		
5.4.2 Evaluation of Individual Verification Results		
a) Does the Food Safety Team systematically evaluate the individual results of planned verification?		
b) If verification does not demonstrate conformity with the planned arrangements, does the Company take action to achieve the required conformity?		
c) Does action to achieve the required conformity include a review of:		
• Existing procedures?		
• Communication channels?		
• The conclusions of the hazard analysis, the established operational PRPs, and the HACCP plan?		
• PRPs?		
• Effectiveness of Human Resource management and training activities?		
5.4.3 Analysis of Results of Verification Activities		
a) Is there evidence to indicate the results of verification activities, including the results of internal and external audits, are analyzed?		
b) What evidence exists to demonstrate that the analysis:		
• Confirms that the overall performance of the system meets the planned arrangements and the FSMS requirements established by the Company?		
• Identifies the need for updating or improving the FSMS?		

ISO 22000 REQUIREMENT	STATUS	ACTION/COMMENTS
• Identifies trends which indicate a higher incidence of potentially unsafe products?		
• Establishes information for planning of the internal audit program concerning the status and importance of areas to be audited?		
• Provides evidence that any corrections and corrective actions that have been taken are effective?		
c) Are results of analyses and resulting activities being recorded and reported in an appropriate manner to Top Management *as input to management reviews*?		
d) Are the results of analyses used as input for updating the FSMS?		
5.5 IMPROVEMENT		
5.5.1 Continual Improvement – Is there evidence indicating that Top Management is committed to continual improvement of the FSMS by means of:		
a) Communication?		
b) Management review?		
c) Internal audits?		
d) Evaluation of individual verification results?		
e) Analysis of results of verification activities?		
f) Validation of control measure combinations?		
g) Corrective actions?		
h) FSMS updating?		
5.5.2 Updating the Food Safety Management System		
a) Is there evidence that Top Management is ensuring that the FSMS is continually updated?		
b) Does a Food Safety Team evaluate the FSMS at planned intervals?		
c) What evidence indicates how the Food Safety Team decides whether it is necessary to review the hazard analysis, the established operational PRP(s), and the HACCP plan?		
d) Are evaluation and updating activities based on:		

ISO 22000 REQUIREMENT	STATUS	ACTION/COMMENTS
• Input from external and internal communications, as stated in 5.6?		
• Input from other sources of information concerning the suitability, adequacy, and effectiveness of the FSMS?		
• Output from the analysis of results of verification activities?		
• Output from management reviews?		
e) Are FSMS updating activities recorded and reported in an appropriate manner (e.g., for input to management reviews)?		

[This page intentionally left blank.]

FS1160-4 – AUDIT REPORT EXAMPLE

Audit Number: Our Company 06-03

Date: July 12, 2006

Purpose: To evaluate the effectiveness of the HACCP Plan and supporting programs

Audit Scope:
- Prerequisite Programs
- Hazard Analysis Preparation
- Hazard Analysis
- HACCP Plan Management

Audit Team:
- Bill Oakes Lead Auditor
- Tom Jenkins Auditor
- Sandra Okumi Auditor
- Clarence Thompson Auditor

Criteria:
- ISO 22000:2005
- Our Company Food Safety Management System

Objective: To determine readiness for annual Food Safety Surveillance audit

Date of Audit: July 7, 2006

Area Personnel:

Rob Portman	Manufacturing Manager
Nancy Givens	Food Safety Manager
Brenda Croft	Product Development
Dean Phelps	Food Safety Lab Supervisor
George Mencke	Quality Manager
Ann Thomas	Purchasing Manager
Mohammad Chowdhury	Process Engineer

Overall Finding

We have conducted an audit of Prerequisite Programs, Hazard Analysis Preparation, Hazard Analyses, and HACCP Plan Management processes and conclude they comply with ISO 22000:2005 and Our Company Food Safety Management System.

Strengths

The HACCP Plan and its supporting systems are well documented, thorough, well communicated, and generally well followed. Evidence shows that a continuing effort is being made to improve the processes and customer satisfaction.

Weaknesses

The practice of using non approved suppliers when materials are not available from approved suppliers should be corrected by expanding the list of approved suppliers or providing approval procedures for bypassing the approved supplier list.

Some records regarding product recalls were found to be lacking in some details.

Obstacles Encountered

None

Number of Nonconformities (NCR's)

Two minor, reference CAR no's: 06-03001 & 06-03002

Observations

The number of sinks for employee hand washing appears to be inadequate for the number of employees needing to wash hands at each shift change.

Recommendations

- Expand the approved supplier list for critical materials
- Install additional hand-washing sinks

Follow-up

We propose a partial re-audit within the next three months to verify corrective action has been taken and is effectively reducing the likelihood of food safety hazards being introduced.

Lead Auditor Signature: ______________________________________
Department Management: ______________________________________
Food Safety Manager: ______________________________________
Date: ____________________

This report is based on random samples; therefore, not every aspect of the Company's activities has necessarily been assessed. Hence, where no nonconformities are reported, it does not follow that none exist.

Doc #: **FS1170**	Title: **CORRECTIVE ACTION**	Print Date: **2/1/2006**
Rev #: **0.0**	Prepared By:	Date: **2/1/2006**
Eff. Date: **2/1/2006**	Reviewed By:	Date Reviewed: **2/1/2006**
	Approved By:	Date Approved: **2/1/2006**
Standards: **ISO 22000:2005, clauses 7.10.1, 7.10.2**		

Policy: To ensure the safety of its products, the Company shall promptly take corrective action whenever it appears food safety is being compromised.

Purpose: This procedure outlines the responsibilities and methods for identifying causes of nonconformities, initiating corrective action(s), and performing follow-up to ensure that the corrective action(s) have been effective in preventing the reason for the nonconformance.

Scope: This procedure applies to all causes of nonconformities relating to product, process, and food safety discovered during production, post-sale, or during internal quality audits.

Responsibilities:

The Food Safety Team Leader (FSTL) is responsible for reporting on corrections and corrective actions taken at Management Review meetings and ensuring that this procedure is accurate, understood, and implemented effectively.

All Company personnel are responsible for identifying nonconforming conditions and initiating a Corrective Action, investigating and recording the cause of nonconforming conditions when assigned, and implementing the corrective actions determined by this procedure.

Definitions: CAR – Corrective action request.

NCR – Nonconformity (or nonconformance) report.

Nonconformity – Object or condition found not conforming to a specific standard or specification (statutory/regulatory, industry, customer, or Company); something that falls outside of identified critical limits. The word "deviation" is sometimes used in place of "nonconformity".

Procedure:

1.0 NONCONFORMITY REPORTS – GENERAL

1.1 Any employee observing an instance of failure of product or process to comply with customer requirements, FSMS standards, or statutory/regulatory requirements should result in the issuance of a FS1170-1 – NONCONFORMITY REPORT.

1.2 Nonconformity reports may be initiated by anyone in the organization and are to be signed off by the manager responsible for that area.

1.3 The nonconformity is described on the report as follows:

- State the requirement briefly, including reference numbers where appropriate;
- Describe the condition actually observed and how it deviates from the requirement;
- Check the appropriate box to indicate whether or not a CAR was also initiated; and
- Request that the area manager sign off on the nonconformity.

NOTE: If an employee observes a nonconformity in an area outside of his/her area of responsibility, the employee should report the nonconformity to the area supervisor/manager, who should review the nonconformity report with the supervisor/manager of the affected area and get his/her signature.

1.4 The FS1170-1 – NONCONFORMITY REPORT is routed to the Food Safety Team Leader for analysis and filing. Trends identified through analysis of the NCR file will be reported to the management team for review as those trends are identified (see FS1020 – MANAGEMENT RESPONSIBILITY).

2.0 INITIATING CORRECTIVE ACTION

2.1 When it is determined that a correction or corrective action is required, such action shall begin with a FS1170-2 – CORRECTIVE ACTION REQUEST. Every CAR must include a description of the problem, observation, or nonconformance, as well as when and where it was observed.

Any employee may initiate a corrective action. Furthermore, ***it is every employee's duty to initiate a corrective action when it appears to them that safety of the Company's end product(s) could be compromised.***

2.2 The completed FS1170-1 shall be submitted to the Food Safety Team Leader. The Food Safety Team Leader shall assign a CAR number to the request and track it using FS1170-3 – CORRECTIVE ACTION LOG.

2.3 The Food Safety Team Leader shall forward the FS1170-2 – CORRECTIVE ACTION REQUEST to the manager of the responsible department, who shall identify a person responsible for investigating and/or taking the necessary actions to correct and prevent the recurrence of the problem. The Department Manager shall notify the FSTL of the assignment, and the FSTL shall identify the individual responsible for the investigation on the Corrective Action Request form.

2.4 The Food Safety Team Leader shall provide one copy of the Corrective Action Request form to the individual responsible for the action or determination of the action. The FSTL shall keep a copy of the FS1170-2 in the Open Corrective Action file.

2.5 The Food Safety Team Leader shall maintain the FS1170-3 – CORRECTIVE ACTION LOG, tracking the status of the Corrective Action Request on the log.

3.0 INVESTIGATING THE CAUSE

3.1 The assigned person shall investigate the problem to determine the underlying, or root, cause or causes. Depending on the nature of the situation under investigation, the assigned person may enlist the aid of other personnel or 'ɜpartments to form a team to investigate and address the problem.

In investigating root cause, keep in mind the apparent cause is rarely the root cause. It is often of value to identify the apparent cause, then the contributing causes. Further analysis using this process can then lead to the root cause of the problem.

3.2 The person(s) investigating shall record any observations, measurements, and the results of this investigation on the FS1170-2 – CORRECTIVE ACTION REQUEST.

4.0 TAKING CORRECTIVE ACTION

4.1 Following the investigation of cause, the Department Manager or an authorized delegate shall review the results and consult with the appropriate personnel to determine what corrective action or actions may be taken to eliminate the cause of the problem. Also, based on the investigator's observations and the department manager's experience and judgment, an action deadline shall be set.

4.2 In the event that the Department Manager assigns responsibility for implementing corrections and corrective actions to a specific individual, the Food Safety Team Leader will record the name of the person and the target date for completion on the FS1170-2 – CORRECTIVE ACTION REQUEST.

5.0 PREVENTING RECURRENCE

5.1 Actions will be taken to rectify and mitigate short-term problems, but the Food Safety Team Leader and the affected department manager shall give consideration to preventing the problem from occurring again or similar potential problems from occurring in other areas, in accordance with the applicable PRP (see FS1050 – PREREQUISITE PROGRAMS), the operational PRP (also see FS1050), and/or the affected HACCP Plan (see FS1080 – HACCP PLAN MANAGEMENT). Recommendations for changes to a PRP or the HACCP plan shall be recorded in the appropriate section of FS1170-2 – CORRECTIVE ACTION REQUEST.

5.2 Upon completion, the person responsible for completing the actions will sign the FS1170-2 – CORRECTIVE ACTION REQUEST and return it to the Food Safety Team Leader for review, in accordance with FS1080 – HACCP PLAN MANAGEMENT.

6.0 VERIFICATION AND CLOSURE

6.1 The Food Safety Team Leader will review the corrective actions taken and determine the appropriate follow-up or verification required. The verification will be briefly described in the verification of implementation section of FS1170-2 – CORRECTIVE ACTION REQUEST. To effectively review the corrective action for its effectiveness, some actions may require a one or two month period for the action to be in place. Because of this, Corrective Actions may remain open for a period of time after the action is taken.

6.2 If the Food Safety Team Leader determines that the action has been effective, he/she will sign and date the form and file it in the Closed Corrective Action File with all other signed copies of the same form.

6.3 If it is determined that the actions taken are not effective, a new FS1170-2 – CORRECTIVE ACTION REQUEST will be generated and the new FS1170-2 will be referenced in the Verification of Implementation section of the previous FS1170-2.

Effectiveness Criteria:

- Timeliness of corrective actions.
- Prevention of recurrence.

References:

A. Food Safety Manual:
  - FSM 7.9.1 – Corrections
  - FSM 7.9.2 – Corrective Actions

B. Food Safety Procedures
  - FS1050 – PREREQUISITE PROGRAMS
  - FS1080 – HACCP PLAN MANAGEMENT

Records:

- FS1170-1 – NONCONFORMITY REPORT
- FS1170-2 – CORRECTIVE ACTION REQUEST
- FS1170-3 – CORRECTIVE ACTION LOG

Revision History:

Revision	Date	Description of changes	Requested By
0.0	2/1/2006	Initial Release	

FS1170-1 – NONCONFORMITY REPORT

<table>
<tr><td colspan="3">NCR Number:</td></tr>
<tr><td>Issued To:</td><td>Issued By:</td><td>Issue Date:</td></tr>
<tr><td colspan="3">Requirement (i.e., Why it is a nonconformity)</td></tr>
<tr><td colspan="3">Nonconformity (i.e., What is wrong): Major ☐ Minor ☐</td></tr>
<tr><td colspan="3">Department/Location:
Management:</td></tr>
<tr><td colspan="3">Action Taken:</td></tr>
<tr><td colspan="3">CAR Issued? Y ☐ N ☐
If "Y", CAR #: ________________</td></tr>
<tr><td colspan="3">Department Manager: ________________________________ Date: __________
Plant Manager: ______________________________________ Date: __________</td></tr>
</table>

[This page intentionally left blank]

FS1170-2 – CORRECTIVE ACTION REQUEST

CAR NO._____

TO (ORGANIZATION NAME, ADDRESS, PHONE NO.)	FROM	
REQUEST DATE	REPLY DUE DATE	
PRODUCT NAME	PRODUCT NO.	
INSPECTION REPORT NO. OR NCR NO. (?)	PROGRAM OR PROJECT	
DESCRIPTION OF CONDITION		
APPARENT CAUSE		
ACTUAL CAUSE		
ACTION TAKEN TO PREVENT RECURRENCE		
SIGNATURE	TITLE	DATE

Form FS1170-2 Page 1 of 2

DATE	NOTES

Form FS1170-2 Page 2 of 2

FS1170-3 – CORRECTIVE ACTION LOG

CAR #	DATE	BLDG	LOCATION	DEPT	PROBLEM DESC	PROB CATEGORY	CAR ORIGINA-TOR	DATE REPORTED	CA ASSGD TO	ACTION TAKEN	DATE RESOLVED

[This page intentionally left blank]

Doc #: **FS1180**	Title: **CONTINUAL IMPROVEMENT**	Print Date: **2/1/2006**
Rev #: **0.0**	Prepared By:	Date Prepared: **2/1/2006**
Eff. Date: **2/1/2006**	Reviewed By:	Date Reviewed: **2/1/2006**
	Approved By:	Date Approved: **2/1/2006**
Standards: **ISO 22000:2005, clause 8.5.1**		

Policy: The Company shall ensure continual improvement of the Food Safety Management System (FSMS) through communication, review, audits, verification activities, validation of control measures, corrective actions, and updating of the FSMS.

Purpose: To establish methods and requirements for determining, collecting, and analyzing appropriate data to demonstrate the effectiveness of the food safety management system, and to provide guidance for continual improvement.

Scope: This procedure applies to all departments included in the food safety management system.

Responsibilities:

The Food Safety Team is responsible for collecting, analyzing, and publishing measurement data; for determining the root cause of process and product food safety problems, and for recommending action to resolve those problems.

All Department Managers are responsible for producing and using process monitoring and measuring data to continually improve the Company's Food Safety Management System. Department Managers are also responsible for implementing the Food Safety Team's recommendations for improvement.

The Management Team (consisting of affected Department Managers and at least one representative of Top Management) is responsible for reviewing QA summaries of data analysis and recommending improvements to the system(s) or process(es) under review.

Definitions: Special Cause Variability – Uncontrolled deviation from the normal distribution of data or events which would otherwise be expected from the system (e.g., failing to maintain/recalibrate production equipment often results in increased product variability). Special causes of variability may be identified, analyzed, and eliminated to bring the system back into control.

Common Cause Variability – Normal distribution of data or events, which is a function of system design. Common cause variability may only be reduced through re-engineering the system. Common cause variability may be reduced, but not eliminated.

Process Capability – Comparison of the normal distribution of data or events with the customer requirement for control of such variation (Specification Limits). A process capability of 1.0 indicates that the process three sigma control limits are equal to the customer specification limits. A process capability of 2.0 indicates that the process three sigma control limits are half of the customer specification limits. This condition, known as *six-sigma capability*, means fewer than 3.4 errors per million attempts.

Deming Cycle – The process-based improvement cycle, also known as the "Plan-Do-Check-Act" cycle, used or referred to in various ISO standards.

Procedure:

1.0 DATA COLLECTION

1.1 The following data, reported by the appropriate departments, shall be collected by the Food Safety Team in accordance with FS1140 – CONTROL OF MONITORING AND MEASURING:

- Food safety nonconformities;
- Product quality and safety data regarding conformity to requirements;
- Process monitoring and measurement data; and
- Supplier performance.

1.2 The information reported shall include, at a minimum:

- Trends and issues related to food safety;
- Product conformity to requirements;
- Characteristics and trends of processes and products, including opportunities for preventive action (i.e., revising PRPs and/or the HACCP plan) and continual improvement; and
- Supplier performance trends and issues.

1.3 The Food Safety Team shall report the information generated from data analyses to the Management Team in a format that quickly shows the important information for management decisions. The information may be provided in a graphic format, with areas of interest or concern highlighted.

2.0 DATA ANALYSIS

2.1 The Food safety Team shall analyze the data using appropriate statistical techniques. To perform adequate and informative analyses, the Food Safety Team shall be adequately trained in the use of statistical analysis tools and techniques, in accordance with FS1030 – COMPETENCE, AWARENESS, AND TRAINING.

2.2 Information regarding process variability as it relates to food safety will be charted visually using appropriate means (e.g., paper charts, computer printouts) to highlight performance trends, effects of process improvement projects, and process problems.

FS1180-1 – VARIABLES CONTROL CHART is one type of control chart suitable for recording and charting process performance data. Other types of charts may be used, as appropriate.

2.3 Any out-of-control condition should result in initiation of an FS1150-1 – NONCONFORMANCE REPORT, in accordance with FS1150 – CONTROL OF POTENTIALLY UNSAFE FOOD PRODUCT. The Food Safety Team Leader shall file nonconformance reports, or NCR's, and periodically review them in accordance with FS1020 – MANAGEMENT RESPONSIBILITY.

2.4 Ongoing problems identified with statistical tools (such as FS1180-1 – VARIABLES CONTROL CHART) shall be subject to analysis, to discover root causes of the problems. Identifying and correcting root causes are central activities in continual improvement.

3.0 DATA REVIEW

3.1 The Management Team shall review data analyses from the Food Safety Team during management review meetings, in accordance with FS1020 – MANAGEMENT RESPONSIBILITY. The purpose of this phase of the management review process is to analyze the variability of the process and product, using statistical indicators to identify opportunities to reduce or eliminate food safety hazards.

3.2 The Management Team shall consider the results of data analyses, control charts, the Company food safety policy, food safety objectives, internal audit results, nonconformity reports, and corrections / corrective actions to continually improve the effectiveness of the Food Safety Management System.

3.3 When specific opportunities are identified by the Management Team, it shall instruct the Food Safety Team to investigate, analyze, and report its findings and recommendations for improvement, in accordance with FS1170 – CORRECTIVE ACTION.

3.4 The Management Team shall review the reports and presentations made by the Food Safety Team and shall approve or decline the recommended actions based on risk, cost and projected return on the recommendations.

If the recommendations are approved, Top Management shall provide the resources required to implement the recommended actions.

3.5 The effectiveness of these actions shall be monitored, measured, and analyzed for future management reviews, in accordance with FS1140 – CONTROL OF MONITORING AND MEASURING.

4.0 DESIGN OF EXPERIMENTS

4.1 The Deming Cycle of process improvement should be the model for implementation of process improvement projects. Prior to implementation, a statistically valid experimental design scheme should be developed and the measuring methods and means established.

4.2 When possible, process improvement projects will be tested prior to full implementation by applying the improved methods to a small portion of the process.

4.3 After implementation and the collection of statistically significant data, an analysis of the data using techniques such as ANOVA (Analysis of Variance) will be conducted. This analysis will establish the magnitude and direction of the system response to the change.

4.4 Based on the analysis, the decision will be made to:

- Fully implement the improvement;
- Conduct further research and/or development; or
- Reject the project and return the test site to prior conditions.

Effectiveness Criteria:

- Reduced process variability
- Continually declining food hazard levels
- Improved food safety processes.
- Fewer deviations from prescribed critical limits.
- Fewer product recalls.

References:

A. Food Safety Manual:
- FSM 5.0 – MANAGEMENT RESPONSIBILITY
- FSM 8.4 – FOOD SAFETY MANAGEMENT SYSTEM VERIFICATION
- FSM 8.5 – CONTINUAL IMPROVEMENT

B. Food Safety Procedures
- FS1020 – MANAGEMENT RESPONSIBILITY
- FS1140 – CONTROL OF MONITORING AND MEASURING

- FS1170 – CORRECTIVE ACTION

Records:

- FS1180-1 – VARIABLES CONTROL CHART
- FS1150-1 – NONCONFORMANCE REPORT
- Other Control Charts
- Data Analysis Reports
- Management Team Meeting Minutes and Reports

Revision History:

Revision	Date	Description of changes	Requested By
0.0	2/1/2006	Initial Release	

[This page intentionally left blank.]

FS1180-1 – VARIABLES CONTROL CHART

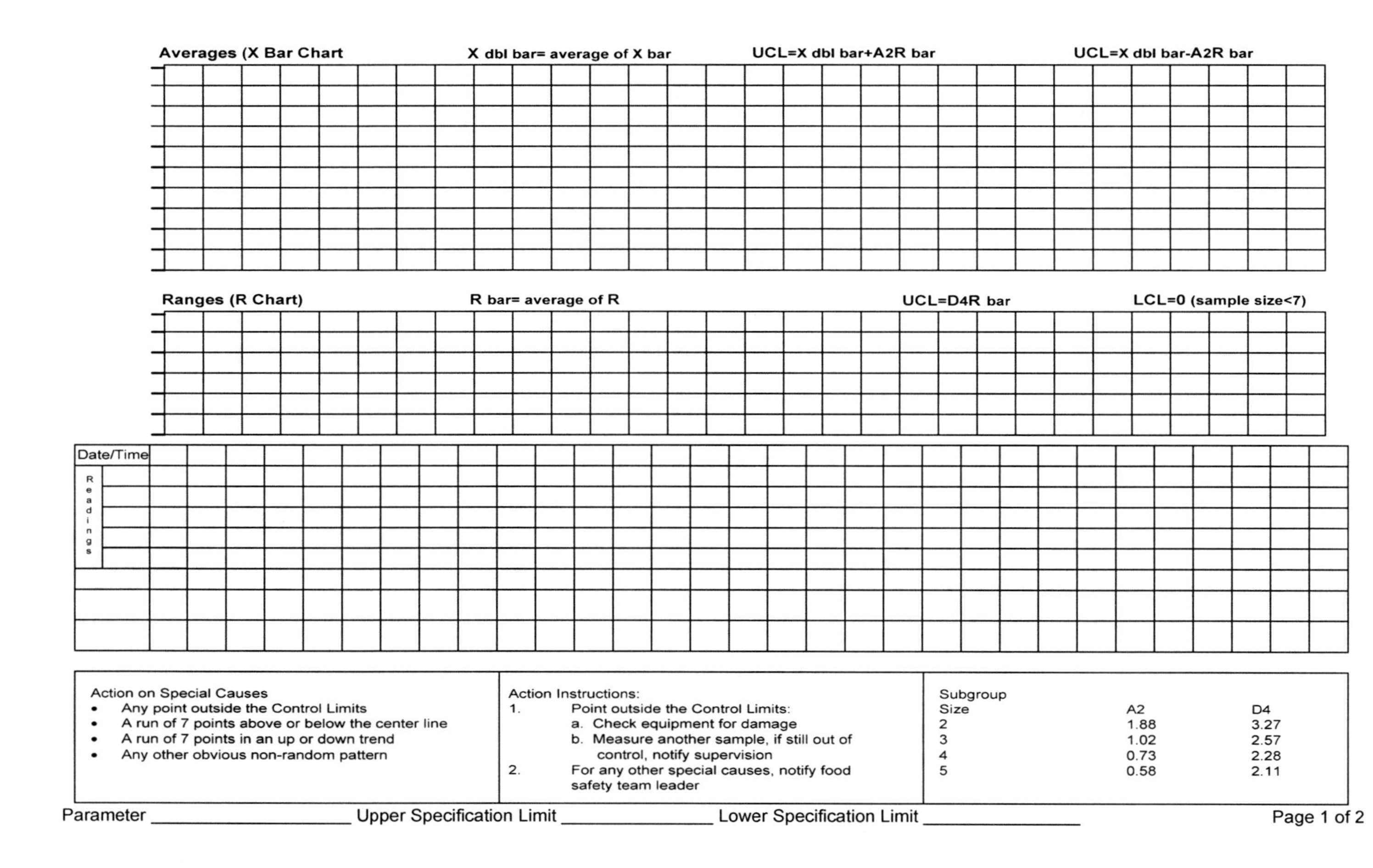

Action on Special Causes	Action Instructions:	Subgroup Size	A2	D4
• Any point outside the Control Limits • A run of 7 points above or below the center line • A run of 7 points in an up or down trend • Any other obvious non-random pattern	1. Point outside the Control Limits: a. Check equipment for damage b. Measure another sample, if still out of control, notify supervision 2. For any other special causes, notify food safety team leader	2 3 4 5	1.88 1.02 0.73 0.58	3.27 2.57 2.28 2.11

Parameter ____________ Upper Specification Limit ____________ Lower Specification Limit ____________ Page 1 of 2

FS1180-1 – VARIABLES CONTROL CHART

Process Log Sheet

ANY CHANGE in the process should be noted

Date	Time	Comments		Date	Time	Comments

Page 2 of 2

Doc #: **FS1190**	Title: **PRODUCT RECALL**	Print Date: **2/1/2006**
Rev #: **0.0**	Prepared By:	Date Prepared: **2/1/2006**
Eff. Date: **2/1/2006**	Reviewed By:	Date Reviewed: **2/1/2006**
	Approved By:	Date Approved: **2/1/2006**
Standards: **ISO 22000:2005, clause 7.10.4**		

Policy: The Company shall immediately and decisively act on all product recalls in order to prevent or minimize harm to the consumer.

Purpose: To outline the procedure for responding to a recall of potentially unsafe Company product; to prevent or minimize the possible harm to the consumer from the recalled product.

Scope: This procedure applies to all recalled products and their associated processes.

Responsibilities:

The Food Safety Team Leader with the food safety team is responsible for initiating and overseeing product recalls and ensuring their adequate implementation and effectiveness.

Quality Assurance (QA) is responsible for determining the scope of the recall and segregation and testing of recalled product to determine proper disposition.

Public Relations (PR) is responsible for communicating information on Company product recalls to regulatory authorities and to the customer/consumer.

All production personnel are responsible for identifying nonconformities which could result in product recalls and for cooperating with the Food Safety Team Leader in the recall effort.

Accounting is responsible for documenting reimbursable costs connected with any product recall and for reimbursing such costs, where applicable.

Definitions: Hold – Time period used for investigation after a food has been identified as potentially unsafe.

Recall – Remove a food product from the market because it may cause health problems or possible death. Food manufacturers or distributors typically issue recalls, which may be based on internal or external findings (e.g., consumer complaints). A product recall may also be known as a "product withdrawal".

Segregate – Remove (recalled) product to an area of storage that spatially isolates it from other foods.

Procedure:

1.0 PRODUCT RECALL INITIATION

1.1 The Food Safety Team Leader will initiate a product recall when a nonconformity having potential food safety consequences is identified. Such nonconformity may be identified through internal control measures, customer complaints, or regulatory requests.

1.2 The Food Safety Team Leader, with possible input from Manufacturing and Quality Assurance, will fill out a 1200-1 – PRODUCT RECALL REQUEST. If all the information indicated on the form is not available, the FSTL will complete the form with the available information.

1.3 The Food Safety Team Leader and Quality Assurance will classify the recall in accordance with FDA/USDA guidelines:

- **Class 1** Health hazard situation with a reasonable likelihood that consuming the product will cause serious health problems or death.
- **Class 2** Situation with a remote likelihood of health problems caused by consuming the product.
- **Class 3** Situation in which product consumption, however unpleasant, will not cause health problems (e.g., salt mistakenly used instead of sugar in hard candy).

1.4 The Food Safety Team Leader will obtain Senior (Top) Management approval for Class 2 and Class 3 recalls. Recalls assessed as Class 1 may be approved on the sole authority of the Food Safety Team Leader if Senior Management is unavailable at the time approval is needed.

2.0 PRODUCT RECALL COMMUNICATIONS

2.1 The approved Product Recall Request shall be distributed to the following departments, at a minimum:

- Quality Assurance;
- Manufacturing;
- Sales;
- Public Relations;
- Shipping/Receiving; and
- Accounting

2.2 The Public Relations Department will notify the FDA District Recall Coordinator, providing all the information available on the Product Recall Request.

2.3 The Public Relations Department, alone or in cooperation with the FDA, will issue a press release and other communications appropriate to the classification of the recall. In the case of a Class 1 recall, timeliness is of the essence.

2.4 Public Relations, with the cooperation with sales and marketing, will identify all customers potentially affected by the recall and notify them in a timely manner, using all available and reasonable contact methods (TV, radio, Company web site/portal, e-mails and phone calls to direct customers, etc.). Sufficient details of the recall shall be given to enable affected parties to take appropriate action.

3.0 HANDLING RECALLED PRODUCT

3.1 The Food Safety Team Leader shall ensure that recalled product can be located, supervise any inventory counts, and account for all recalled food (see FS1130 – IDENTIFICATION, LABELING, AND TRACEABILITY).

3.2 Recalled product shall be handled in compliance with FS1150 – CONTROL OF NONCONFORMING PRODUCT. The Food Safety Team Leader shall ensure the segregation and securing of any recalled product as it is returned.

3.3 Quality Assurance will test recalled product, as required, and submit test results to the Food Safety Team Leader to assure the proper disposition of the affected product.

3.4 The Food Safety Team Leader shall ensure that all records, documents, and files related to the product recall are maintained for a period of no less than three (3) years from the initial recall date.

3.5 Accounting shall ensure that reimbursable costs are properly documented and are reimbursed as soon as possible.

4.0 REVIEW OF PRODUCT RECALL PROCESS

4.1 The Food Safety Team shall audit (annually, at a minimum) logs and other records related to product recalls, to determine if recall procedures are being properly implemented and continue to meet statutory/regulatory requirements.

4.2 The Food Safety Team shall periodically test recall procedures (at least annually), to determine the capability of the procedures and to verify the currency of information contained in them. Records of recall procedure tests will be maintained for at least three years.

4.3 A third-party audit of the recall process should be conducted at regular intervals to verify that the process is properly documented and is communicated to all employees, that it is (capable of being) implemented consistently, meets the necessary requirements, and is monitored and measured.

Effectiveness Criteria:

- Recall completed in fewer than ten (10) calendar days.
- Product recalled before consumers harmed by it.

Additional Resources:

A. How to Report Problems With Products Regulated by the FDA – http://www.fda.gov/opacom/backgrounders/problem.html

B. Responding to a Food Recall, National Food Service Management Institute (NFSMI), in conjunction with the USDA Food & Nutrition Service. See http://www.nfsmi.org.

References:

A. Food Safety Manual
   - FSM 7.9.4 – Recalls (Withdrawals)

B. Food Safety Procedures
   - FS1150 – CONTROL OF NONCONFORMING PRODUCT
   - FS1170 – CORRECTIVE ACTION

Records:

- FS1190-1 – FOOD RECALL CHECKLIST
- Copies of all communications related to recalls

Revision History:

Revision	Date	Description of changes	Requested By
0.0	2/1/2006	Initial Release	

FS1190-1 – FOOD RECALL CHECKLIST

Use this Checklist when carrying out a food recall. The Checklist identifies responsibilities at the administrative (part 1) and operational site level (part 2).

Part 1: Responsibilities at the Administrative (Managerial) Level

TASK	PERSON RESPONSIBLE	COMPLETION DATE
Have a copy of the Company's procedure related to food recalls.		
Write the food recall notice and review with senior management. • Describe the problem; • Determine actions that must be taken. • Review specific directions in the communication(s).		
Communicate the food recall notice to all Company sites • Document that the recall notice was received at each site.		
Collect health-related information needed for public communications for FDA Class 1 or Class 2 recalls[1]. The following information must be collected and documented: • Determination of where the product was served, to whom, and when; and • Reports of health problems potentially related to the recalled product (if it has been served), including physical symptoms of illness and any actions taken.		
Provide the following to Public Relations: • Copies of the food recall notice; • Any additional information related to the recall; • Information on whether the product has been used and served, including to whom and the date of service; and • Reports of potential adverse health risks related to the recalled product.		
• Identify and record whether any product was received by customers. • Locate the recalled food product by site. • Verify that the food item bears product identification code(s) and production date(s) listed in the recall notice.		
Obtain accurate inventory counts of the recalled product from every customer, including amount in inventory and amount used.		
Account for all recalled food by verifying inventory counts against records of food received at customer sites.		
Confirm that customer sites have segregated and secured the recalled food product.		

[1] **FDA Recall Class 1** – Health hazard situation with a reasonable likelihood that consuming the product will cause serious health problems or death. **Class 2** – Situation with a remote likelihood of health problems caused by consuming the product. **Class 3** – Situation in which product consumption, however unpleasant, will not cause health problems (e.g., salt mistakenly used instead of sugar in hard candy).

TASK	PERSON RESPONSIBLE	COMPLETION DATE
Carry out the following administrative tasks as appropriate: • Determine if the food is to be returned (to whom), or destroyed (by whom). • Notify site personnel of procedures, dates, etc. to be followed for collection or destruction of food product. • If the product is to be returned, follow procedures. The product should be consolidated for collection as soon as possible, but no later than 30 days after the date of the recall notification. • If the product is to be destroyed, confirm that written notification is on file. Follow procedures provided by State and local agencies. • Consolidate documentation from all sites for inventory counts. • Document any reimbursable costs.		
Complete and maintain all required documentation related to the recall, such as: • Recall notice and communications about the product; • Records of how food product was returned or destroyed; and • Reimbursable costs		
Maintain copies of all communications received or sent in relation to the food recall for three (3) years plus the current year.		
Maintain copies of all information provided to public communication contact person, other media contact(s), and the public; adverse health reports and actions taken for three years plus the current year.		

Part II: Responsibilities at the Operational (Site) Level

TASK	PERSON RESPONSIBLE	COMPLETION DATE
When the recall notice is received, identify the recalled food product immediately using the product code(s), lot number(s), and date(s) of production run.		
Hold the product using the following steps: • Physically segregate the product, including any open containers, leftover product, and food items in current production that contain the recalled food. If an item is suspected to contain the product but label information is not available, follow the Company's procedure for proper disposal. • Clearly and prominently mark the product "ON HOLD", "DO NOT USE", and "DO NOT DISCARD." • Inform the entire staff the marked product is not to be used. • Document the quantity in inventory.		
Determine if the product has been used by reviewing invoices, production records, inventory records, and menu records.		
Make sure suspect product has been removed from use. • Account for all product received by customers.		

TASK	PERSON RESPONSIBLE	COMPLETION DATE
• Add the amount of product in inventory to the amount already used – that sum *should* equal the amount received.		
If recalled product has been used, document: • Date(s) product used; and • To whom (by class, not by individual) product was served.		
If you have any reports of health problems that could be related to consumption of the food product, direct anyone affected to appropriate medical personnel. Collect information to submit to the main Company office, including persons' names, reported symptoms of physical illness, and actions taken.		
Submit to the Company's Food Safety Team Leader: • Inventory counts of the recalled product and counts or amounts of the product used before the recall notice was received; • How the recalled product was segregated and secured to prevent further use; • Information on whether the product was served; if served, to whom (classes) and the dates of service; • Reports of health problems and actions taken.		
Follow Food Safety's instructions regarding collection, return, and disposition of recalled product.		
Complete any necessary documentation for collection, return or destruction, and reimbursement.		
Submit necessary documentation to the Company's main office, to the attention of the primary Food Safety official.		
Maintain copies of documentation on file for three (3) years plus the current year, which should include (at a minimum): • Copies of communications received regarding the food recall; and • Documentation related to the food recall that demonstrates required procedures were followed, including: a. How the product was secured to prevent its use; b. Return of the product to the distributor/manufacturer or the central office/warehouse; c. Destruction of the product on site, if so notified; d. Records showing the date on which the product was used and to whom it was served; and e. Reports from customers of symptoms of physical illness.		

[This page intentionally left blank]

Document #: **FS1200**	Title: **EMERGENCY PREPAREDNESS AND RESPONSE**	Print Date: **2/1/2006**
Revision #: **0.0**	Prepared By:	Date Prepared: **2/1/2006**
Effective Date: **2/1/2006**	Reviewed By:	Date Reviewed: **2/1/2006**
	Approved By:	Date Approved: **2/1/2006**
Standards: **ISO 22000:2005, clause 5.7**		

Policy: Top management shall establish, implement and maintain procedures to manage potential emergency situations and accidents that can impact food safety and which are relevant to the organization in the food chain.

Purpose: To define the methods for effectively planning for, and acting during, emergencies that actually or potentially impact food safety and other company operations.

Scope: This procedure covers all activities required to prepare for and respond to any emergency. Emergency situations include external and internal causes, both accidental and deliberate. It applies to all management and staff personnel as indicated.

Responsibilities:

The Emergency Management Team is responsible for developing the Emergency Response Plan

The Food Safety Team provides inputs to the plan regarding potential food safety issues.

Managers and Supervisors are responsible for directing emergency response activities in their respective departments.

All personnel are responsible for knowing and understanding the emergency response plan as it relates to their activities and food safety.

Definition: Emergency – Event or circumstance that may significantly impact operations. This includes any sudden, unexpected occurrence, natural or human-caused, which poses a significant threat to food safety and/or the company's personnel, customers, facilities, assets, records, and business activities.

Procedure:

1.0 RISK ASSESSMENT AND EVALUATION

1.1 The Food Safety Team shall periodically (annually, at a minimum) evaluate potential risks to food safety due to emergency occurrences. This type of evaluation may or may not be included in a hazard analysis (see FS1070 – HAZARD ANALYSIS).

1.2 Food safety risks due to emergency situations shall be assessed and prioritized on the basis of their probability and seriousness. The Food Safety Team shall forward written results and recommendations for preventive and/or contingency actions to the Emergency Management Team for review and incorporation into the Emergency Response Plan.

1.3 The Food Safety Team will use the following criteria:

- Any potential risk shall be identified, whether already addressed or newly anticipated. If the identified risk has already been successfully addressed, no further assessment or recommendations shall be required. Any newly-identified risk shall be fully addressed.
- Identified risks may be categorized as natural or caused by human action and the latter may be broken down into accidental and intentional. Situations that have the potential for creating an adverse impact on the company's operations and food safety may include (but are not necessarily limited to) the following:

  a. Fire;
  b. Power outage;
  c. Water leak;
  d. Gas leak;
  e. Bomb threat;
  f. Earthquake;
  g. Flood;
  h. Toxic (hazardous) waste dispersion;
  i. Civil disturbance;
  j. Bomb explosion;
  k. Sabotage;
  l. Bioterrorism;
  m. Tornado;
  n. Hurricane; and
  o. Nuclear attack.

1.4 Each potential risk shall be evaluated according to the following guidelines:

- What is the Company's ability to predict this occurrence?
- What is the likelihood of this occurrence?
- If applicable, what is the estimated elapsed time between your prediction and the actual onset of this occurrence?
- If applicable, what is the estimated elapsed time between discovery of the emergency and any impact to the safety of food products?
- What is the possibility of controlling or diminishing the impact of this occurrence?

- What is the anticipated duration of this emergency?
- What is the projected impact to the Company regarding:
  a. Product safety?
  b. Personnel injury or working conditions?
  c. Customer confidence and the Company's end product(s)?
  d. Security measures necessary to ensure adequate protection of personnel, facilities, assets, and food safety records?
  e. Legal issues surrounding the Company's ability to control food safety hazards during and after the emergency?
  f. Requirements to backup or substitute data or support services?
  g. The emergency's effect upon related functions?
  h. Time for the facility (and Company) to recover fully from the emergency?
  i. Losses due to spoilage or contamination?
  j. Equipment lost?

1.5 Once potential risks have been identified and prioritized, the Emergency Management Team shall:

- Determine potential short- and long-range solutions;
- Decide upon the most appropriate primary and secondary solutions;
- Test and evaluate these solutions for effectiveness and document such tests on FS1200-1 – RISK MANAGEMENT SOLUTIONS TEST REPORT; and
- Include these solutions in the Emergency Response Plan and related testing and training programs.

1.6 All original records of the processes used to compile the information resulting in these policies and procedures shall be retained by the Emergency Management Team Coordinators, in a location separate from the Emergency Response Plan. These records shall include the names and positions of all appropriate persons involved in the continued planning and writing of this plan.

1.7 The Emergency Management Team will review all insurance policies and service and vendor agreements with regard to their provisions for emergency response at least annually.

- If policies and agreements are satisfactory, existing agreements shall be renewed. Problem policies and agreements shall be renegotiated, as necessary.

2.0 EMERGENCY RESPONSE PLANNING

2.1 The following information, at a minimum, should be incorporated into the FS1200-2 – EMERGENCY RESPONSE PLAN:

- Emergency services contact information and protocols;
- Emergency phone numbers, report forms, and emergency instructions;

- Description of emergency warning devices and signals, such as sirens, air horns, and alarm lights;
- Facility evacuation plans, assembly points, and rosters of employees assigned to each assembly point;
- Designation of responsibility for evacuation of all employees, visitors, and others to a department manager and/or designees;
- Placement of fire extinguishers, emergency exits, gas and water valves, electrical power shutoff devices, first aid kits, and instructions for the safe and effective operation of each device;
- Location of containment facilities and location marking materials for control of affected product;
- Procedures for emergency operation, shutdown, and securing of critical processes and associated equipment;
- Procedures for recovery and start-up of critical processes and equipment after an emergency;
- Procedures for containment and clean-up of hazardous material spills;
- List of critical data and records to be secured and directions for securing them;
- List of personnel to be contacted and criteria for making contact with each;
- Instructions and record forms for conducting emergency drills;
- Emergency procedures training records; and
- Other issues which may be identified in the course of hazard analyses.

2.2 Activities conducted in the course of responding to an actual emergency or testing the FS1200-2 – EMERGENCY RESPONSE PLAN shall be recorded on FS1200-3 – EMERGENCY RESPONSE ACTIVITY LOG.

2.3 The Food Safety Team Leader shall maintain the FS1200-2 – EMERGENCY RESPONSE PLAN and any FS1200-3 – EMERGENCY RESPONSE ACTIVITY LOG related to food safety emergency response or testing, in accordance with FS1010 – FOOD SAFETY RECORDS.

3.0 RESPONDING TO AN EMERGENCY

3.1 During an emergency:

- Remain calm and avoid any actions that might jeopardize the safety of employees, customers, visitors, or other persons or the integrity of the Company;
- Activate the FS1200-2 – EMERGENCY RESPONSE PLAN;
- Initiate and maintain an FS1200-3 – EMERGENCY RESPONSE ACTIVITY LOG, if at all possible; and

- Assist in controlling the impact of the emergency on employees, visitors, and other personnel and Company property according to the FS1200-2 – EMERGENCY RESPONSE PLAN.

3.2 After the emergency:

- Assess and report the impact of the emergency to the Emergency Management Team Leader;
- Carry out containment, clean-up, recovery, and start-up procedures as defined in the Emergency Response Plan;
- Refer all requests for interviews or other information to the Emergency Management Team Leader; and
- Any employee keeping a FS1200-3 – EMERGENCY RESPONSE ACTIVITY LOG shall forward it to the Emergency Management Team for review.

3.3 Within two weeks of the emergency, the Emergency Management Team shall review all FS1200-3 – EMERGENCY RESPONSE ACTIVITY LOG with the Food Safety Team, to determine if the FS1200-2 – EMERGENCY RESPONSE PLAN was adequate or if it requires revision. Any modification to the Plan should be tested at the earliest possible date to ensure its adequacy (see section 4.0).

4.0 EMERGENCY DRILLS AND TESTS

4.1 The Emergency Management Team will conduct tests whenever a new risk is identified and solutions for prevention or mitigation developed. A FS1200-1 – RISK MANAGEMENT SOLUTIONS TEST REPORT will be used to describe the test scenario and solutions and to record the results of the tests.

4.2 The Emergency Management Team will conduct quarterly drills to test various aspects of the FS1200-2 – EMERGENCY RESPONSE PLAN. Test activity will include both scheduled and unannounced drills.

4.3 Results of quarterly drills will be recorded on an FS1200-3 – EMERGENCY RESPONSE ACTIVITY LOG. This log shall be completed in a chronological sequence of events and clearly identify the following information:

- Time each activity began and all actions taken;
- Location(s) involved in drill/test;
- Person(s) maintaining log entries;
- Person(s) generating additional information;
- Names/titles of persons assigned to specific tasks;
- Notification of other personnel, if applicable:
  a. Name of person notified;
  b. Time of notification;

c. Action requested; and
d. Disposition.

- Unusual events requiring follow-up when normal operations resume; and
- Termination date/time of emergency activation, including name and title of person terminating drill/test.

4.4 Response Plan deficiencies identified through testing and drills shall result in modifications to the plan. Modifications to the FS1200-2 – EMERGENCY RESPONSE PLAN should be tested at the earliest possible date to ensure their adequacy.

Effectiveness Criteria:

- No unforeseen emergencies.
- In the event of an emergency, employee response is appropriate and injury to persons and property is minimized.
- Employees understand the Emergency response Plan and their responsibilities in regard to the plan.
- Emergency response drills are conducted and records kept.

References:

A. Food Safety Manual
   - FSM 5.7 – Emergency Preparedness and Response
   - FSM 7.4 – Hazard Analysis

B. Food Safety Procedures
   - FS1010 – Food Safety Records
   - FS1070 – Hazard Analysis

C. Statutory/Regulatory Requirements

   There are numerous federal, state, and local laws pertaining to response and recovery in the event of a food safety emergency. The Company must be aware of and must comply with all applicable laws.

Records:

- FS1200-1 – RISK MANAGEMENT SOLUTIONS TEST REPORT
- FS1200-2 – EMERGENCY RESPONSE PLAN
- FS1200-3 – EMERGENCY RESPONSE ACTIVITY LOG

Revision History:

Revision	Date	Description of changes	Requested By
0.0	2/1/2006	Initial Release	

[This page intentionally left blank.]

FS1200-1 – RISK MANAGEMENT SOLUTIONS TEST REPORT

On ______________________ (date), the Disaster Management Plan solutions to the scenario indicated below were tested with the following results:

RISK:

PROPOSED PREVENTIVE/CONTINGENCY ACTION:

SCENARIO:

Test conducted by: ____________________

Location/Department: ____________________

Date: ____________________ Time: ____________________

RESULTS:

[This page intentionally left blank]

FS1200-2 – EMERGENCY RESPONSE PLAN

1.0 EMERGENCY MANAGEMENT TEAM

The following personnel comprise the Emergency Management Team:

TEAM LEADER: ______________________________

Office Phone: ____________________

Home Phone: ____________________

Cell Phone: ____________________

Pager: ____________________

Responsibilities:

- Public and Media Relations
- Shareholder and Stockholder Relations

Backup: ______________________________

Office Phone: ____________________

Home Phone: ____________________

Cell Phone: ____________________

Pager: ____________________

ADMINISTRATIVE
COORDINATOR: ______________________________

Office Phone: ____________________

Home Phone: ____________________

Cell Phone: ____________________

Pager: ____________________

Responsibilities:

- Personnel
- Insurance
- Accounting
- Data Processing, Administrative Functions

Backup: ______________________________

Office Phone: ____________________

Home Phone: ____________________

Cell Phone: ____________________

Pager: ____________________

OPERATIONS
COORDINATOR: ______________________________

Office Phone: ____________________

Home Phone: ____________________

Cell Phone: ____________________

Pager: ____________________

Responsibilities:

- Operations
- Security
- Telecommunications
- Data Processing, Operations Functions

Backup: ______________________________________

Office Phone: ____________________

Home Phone: ____________________

Cell Phone: ____________________

Pager: ____________________

ENVIRONMENTAL and
SAFETY COORDINATOR: ________________________

Office Phone: ____________________

Home Phone: ____________________

Cell Phone: ____________________

Pager: ____________________

Responsibilities:

- Emergency Services Liaison
- Spill Control and Clean-up
- Regulatory Affairs

Backup: ______________________________________

Office Phone: ____________________

Home Phone: ____________________

Cell Phone: ____________________

Pager: ____________________

FOOD SAFETY
TEAM LEADER: ______________________________

Office Phone: ____________________

Home Phone: ____________________

Cell Phone: ____________________

Pager: ____________________

Responsibilities:

- Customer Liaison
- Product Safety
- Control of Potentially Unsafe Product

Backup: ______________________________________

Office Phone: ____________________

Home Phone: ____________________

Cell Phone: ____________________

Pager: ____________________

PLANT ENGINEER: ______________________________

Office Phone: ____________________

Home Phone: ____________________

Cell Phone: ____________________

Pager: ____________________

Responsibilities:

- Utilities

Backup: ______________________________________

Office Phone: ____________________

Home Phone: ____________________

Cell Phone: ____________________

Pager: ____________________

2.0 EMERGENCY CONTACT INFORMATION

- Plant fire department ______________________________
- Local fire department ______________________________
- Regional FDA coordinator ___________________________
- Regional FEMA Office ______________________________
- Local police ______________________________________
- Plant Physician/Nurse ______________________________
- Hospital ____
- Environmental Authority ____________________________

3.0 NOTIFICATION AND RESPONSE

3.1 GENERAL

Upon notification of an emergency, the Emergency Management Team Administrative or his/her designee shall authorize the notification and immediate response of appropriate personnel. This notification shall be made by phone or in person, and according to one of the three levels of threat as follows:

3.2 ALERT 3: STAND-BY:

Appropriate personnel are to be notified of the emergency and placed on stand-by status, and shall remain available for emergency assignment, either by phone or by other means as necessary. If the emergency occurs during business hours:

- All personnel are placed on alert
- Critical processes and or equipment are prepared for emergency operation and shutdown as defined in the Emergency Response Plan
- All appropriate records, equipment and documents are prepared for transfer to secure areas or facilities.
- All operations may resume routine functions.

3.3 ALERT 2: MINIMUM RESPONSE:

Only selected personnel will be called to respond immediately, and all other personnel will remain on stand-by status. If the emergency occurs during business hours:

- All personnel are to remain at current locations, until instructed to leave.
- All appropriate records, equipment and documents are transferred to secure areas or facilities; and business operations involving customers are discontinued, and internal tasks, functions and assignments continue until further instructions are received.

3.4 ALERT 1: MAXIMUM RESPONSE:

All designated personnel shall respond as assigned.

The Company or any of its facilities may be closed immediately, and all personnel not responsible for dealing with the emergency remain at their current locations, or are evacuated as defined in the Emergency Response Plan.

- All appropriate records, equipment and documents are secured
- All business operations are discontinued. Employees and other personnel remain at identified locations until escorted from those locations by management or emergency services personnel.

3.5 In considering the appropriate response, the Emergency Management Team shall consider which stage(s) of emergency is present:

- Detection
- Reaction
- Assessment
- Notification
- Mobilization
- Recovery
- Resumption

4.0 EMERGENCY WARNINGS

5.0 EVACUATION MAPS

6.0 EMERGENCY PROCEDURES

6.1 Operation, Shut-down and Securing Procedures

- Data and records Security

6.2 Recovery and Start-up Procedures

6.3 Containment and Clean-up Procedures

7.0 EMERGENCY DRILLS

8.0 EMERGENCY TRAINING RECORDS

[This page intentionally left blank]

FS1200-3 – EMERGENCY RESPONSE ACTIVITY LOG

DATE:_______________TIME STARTED: _____________TIME ENDED:_____

LOCATION:____

INITIATOR:____________________________TEAM LEADER:___________

TEST: Y / N

Time Begin	Person	Activity/request	Disposition	Time End

[This page intentionally left blank]

Food Safety Management System Policies, Procedures & Forms

Section 500

Index

Section 500

Index

Index

Document / Section Title	Section ID	Page
178/2002 (Article 18) – European Parliament (EC) General Food Law Regulation	FS1130	267
19011:2002 – ISO/EN Guidelines for Quality and/or Environmental Management Systems Auditing	FS1160	296
21 CFR 101 (USA) – Food Labeling	FS1130	267
21 CFR 110 (USA) – Current Good Manufacturing Practice in Manufacturing, Packing, or Holding Human Food	FS1120	261
22000:2005 – ISO/FDIS Food Safety Management Systems / Requirements for Any Organization in the Food Chain	FS1160	296
852/2004 – European Parliament Regulation (EC), re, Hygiene of Foodstuffs	FS1070	200
854/2004 – European Parliament Regulation (EC), re, Organization of Official Controls on Products of Animal Origin Intended for Human Consumption	FS1110	251
9 CFR 417 (USA) – Hazard Analysis and Critical Control Point (HACCP) Systems	FS1080	212
A		
Americans with Disabilities Act (1990 - USA)	FS1040	165
Approval and Distribution, Job Description	FS1040	166
Approved Chemicals / Authorized Handlers List Example (FS1050-3)	FS1050	181
Approved Vendor List - form (FS1100-1)	FS1100	237
Approved Vendor List – procedure	FS1100	232
Approved Vendor Notification form (FS1100-3)	FS1100	245
Assessment and Evaluation, Risk	FS1200	359
Audit (Internal) Follow-Up	FS1160	296
Audit (Internal) Planning	FS1160	293
Audit (Internal) Process, Validation of	FS1160	296
Audit (Internal) Program	FS1160	292
Audit (Internal) Reporting	FS1160	295
Audit (Vendor) Program, Second-Party	FS1100	235
Audit Plan (FS1160-2)	FS1160	301
Audit Program (FS1160-1)	FS1160	299
Audit Report example (FS1160-4)	FS1160	331
Auditing, Guidelines for Quality and/or Environmental Management Systems (ISO/EN 19011:2002)	FS1160	296
B		
Bioterrorism Act of 2002 (USA)	FS1010	145
Bioterrorism Act of 2002, Regulation 306 (USA)	FS1130	267
C		
Calibration Database (FS1140-2)	FS1140	279
Calibration Process Review	FS1140	275
Calibration Record (FS1140-1)	FS1140	277

Document / Section Title	Section ID	Page
Calibration System	FS1140	273
Canada, Consolidated Regulations of (CRC) – Fish Inspection Regulations c. 802	FS1080	212
Canada Consumer Packaging and Labeling Act (R.S. 1985, c. C-38)	FS1130	267
Canada, Fish Inspection Act	FS1110	251
Characteristics, Product	FS1060	188
Checklist, Food Recall (FS1190-1)	FS1190	355
Checklist, Food Safety Audit (FS1160-3)	FS1160	303
Checklist, Hazard Analysis (FS1070-1)	FS1070	201
Classification and Requirements, Vendor	FS1100	232
Closure, Verification and (Corrective Action)	FS1170	335
Communications, Product Recall	FS1190	352
Conducting the Internal Audit	FS1160	293
Consumer Packaging and Labeling Act (Canada – R.S. 1985, c. C-38)	FS1130	267
Control Chart, Variables (FS1180-1)	FS1180	349
Control Measures, Flow Diagrams, Process Steps	FS1060	189
Control Measures, Selection and Assessment of	FS1070	198
Control of Subcontractor Calibration	FS1140	275
Control Process Review (nonconforming product)	FS1150	284
Controls on Products of Animal Origin Intended for Human Consumption, Organization of Official (EC #854/2004)	FS1110	251
Corrective Action - Investigating the Cause	FS1170	335
Corrective Action - Preventing Recurrence	FS1170	335
Corrective Action - Verification and Closure	FS1170	335
Corrective Action (control of nonconforming product)	FS1150	283
Corrective Action Log (FS1170-3)	FS1170	341
Corrective Action Request form (FS1170-2)	FS1170	339
Corrective Action, Initiating a	FS1170	334
Corrective Action, Taking	FS1170	335
CRC (Consolidated Regulations of Canada) – Fish Inspection Regulations c. 802	FS1080	212
Current Good Manufacturing Practice in Manufacturing, Packing, or Holding Human Food (21 CFR 110, USA)	FS1120	261

D

Document / Section Title	Section ID	Page
Data Analysis	FS1180	345
Data Collection	FS1180	344
Data Review	FS1180	345
Design of Experiments	FS1180	346
Developing a HACCP Plan	FS1080	206
Discrepancies, Disposition, Rejection	FS1110	250
Disposition - Rejection, Discrepancies, and	FS1110	250
Disposition of Potentially Unsafe Products	FS1150	284
Disqualification, Vendor	FS1100	234
Document Change Control (FS1000-2)	FS1000	139
Document Control Database (FS1000-3)	FS1000	141
Document Control Process Review	FS1000	135
Document Distribution	FS1000	134
Document Revision	FS1000	133

Document / Section Title	Section ID	Page

E

Emergency Drills and Tests	FS1200	363
Emergency Response Plan (FS1200-2)	FS1200	369
Emergency Response Activity Log (FS1200-3)	FS1200	375
Emergency Response Planning	FS1200	361
Emergency, Responding to	FS1200	362
Employee Orientation, New	FS1030	156
Employee Selection, New	FS1030	155
Establishing Operational Prerequisite Programs	FS1050	174
European Parliament Regulation (EC) number 178/2002 – General Food Law, Article 18	FS1130	267
European Parliament Regulation (EC) number 852/2004 – Hygiene of Foodstuffs	FS1070	200
European Parliament Regulation (EC) number 854/2004 – Organization of Official Controls on Products of Animal Origin Intended for Human Consumption	FS1110	251
Evaluation, New Vendor	FS1100	233

F

Final Inspection	FS1120	259
Final Release (of product)	FS1120	260
Fish Inspection Act (Canada)	FS1110	251
Fish Inspection Regulations, Consolidated Regulations of Canada (CRC), c. 802	FS1080	212
Flow Diagram Example (FS1060-1)	FS1060	193
Flow Diagrams, Process Steps, and Control Measures	FS1060	189
Food Items and Ingredients, Labeling / Tracing	FS1130	264
Food Labeling (21 CFR 101, USA)	FS1130	267
Food Recall Checklist (FS1190-1)	FS1190	355
Food Safety Audit Checklist (FS1160-3)	FS1160	303
Food Safety Management Systems / Requirements for Any Organization in the Food Chain (ISO/FDIS 22000:2005)	FS1160	296
Food Safety Record Generation	FS1010	144
Food Safety Record Maintenance	FS1010	144
Food Safety Records List (FS1010-1)	FS1010	147
Food Safety Records, Identification of	FS1010	143
Food Safety Team, The	FS1060	188
Food Safety Training Log (FS1030-2)	FS1030	161
Food Safety Training Requirements List (FS1030-1)	FS1030	159
Format and Content, Job Description	FS1040	164
FS1000-1 – Request for Document Change	FS1000	137
FS1000-2 – Document Change Control	FS1000	139
FS1000-3 – Document Control Database	FS1000	141
FS1010-1 – Food Safety Records List	FS1010	147
FS1030-1 – Food Safety Training Requirements List	FS1030	159
FS1030-2 – Food Safety Training Log	FS1030	161

Document / Section Title	Section ID	Page
FS1040-1 – Job Description	FS1040	169
FS1050-1 – Prerequisite Program Example	FS1050	177
FS1050-2 – Standard Operating Procedure (SOP) Form – Example	FS1050	179
FS1050-3 – Approved Chemicals / Authorized Handlers List Example	FS1050	181
FS1050-4 – Storage Map Example	FS1050	183
FS1050-5 – PRP Log Example	FS1050	185
FS1060-1 - Flow Diagram Example	FS1060	193
FS1070-1 - Hazard Analysis Checklist (Example)	FS1070	201
FS1080-1 - HACCP Plan Worksheet	FS1080	215
FS1080-2 - HACCP Plan Outline	FS1080	217
FS1090-1 - Purchase Requisition	FS1090	223
FS1090-2 - Purchase Order	FS1090	225
FS1090-3 - Purchase Order Log	FS1090	227
FS1090-4 - Purchase Order Follow-Up	FS1090	229
FS1100-1 - Approved Vendor List	FS1100	237
FS1100-2 - Vendor Survey Form	FS1100	239
FS1100-3 - Approved Vendor Notification	FS1100	245
FS1100-4 - Vendor Performance Log	FS1100	247
FS1110-1 - Receiving Log	FS1110	253
FS1110-2 - Receiving and Inspection Report	FS1110	255
FS1130-1 - Lot Identification / Product Traceability Log	FS1130	269
FS1140-1 - Calibration Record	FS1140	277
FS1140-2 - Calibration Database	FS1140	279
FS1150-1 - Nonconformance Report	FS1150	287
FS1150-2 - Returned Goods Authorization	FS1150	289
FS1160-1 - Audit Program	FS1160	299
FS1160-2 - Audit Plan	FS1160	301
FS1160-3 - Food Safety Audit Checklist (example)	FS1160	303
FS1160-4 - Audit Report (example)	FS1160	331
FS1170-2 - Corrective Action Request	FS1170	339
FS1170-3 - Corrective Action Log	FS1170	341
FS1170-3 - Nonconformity Report	FS1170	337
FS1180-1 - Variables Control Chart	FS1180	349
FS1190-1 - Food Recall Checklist	FS1190	355
FS1200-1 - Risk Management Solutions Test Report	FS1200	367
FS1200-2 - Emergency Response Plan	FS1200	369
FS1200-3 - Emergency Response Activity Log	FS1200	375
G		
General Information, Hazard Analysis	FS1070	196
Good Manufacturing Practice (Current) in Manufacturing, Packing, or Holding Human Food (21 CFR 110, USA)	FS1120	261
Guidelines for Quality and/or Environmental Management Systems Auditing (ISO/EN 19011:2002)	FS1160	296
H		
HACCP Plan Outline (FS1080-2)	FS1080	217
HACCP Plan Review	FS1080	210

Document / Section Title	Section ID	Page
HACCP Plan Revision	FS1080	211
HACCP Plan Worksheet (FS1080-1)	FS1080	215
HACCP Plan, Developing	FS1080	206
HACCP Plan, Implementing	FS1080	210
Handling Recalled Product	FS1190	353
Handling, Maintenance, Storage (monitoring / measuring equipment)	FS1140	273
Hazard Analysis - General Information	FS1070	196
Hazard Analysis and Critical Control Point (HACCP) Systems – Code of Federal Regulations (USA), Title 9, Chapter 3, Part 417	FS1080	212
Hazard Analysis Checklist - example (FS1070-1)	FS1070	201
Hazard Analysis Preparation – Verification	FS1060	190
Hazard Analysis Review	FS1070	199
Hazard Assessment	FS1070	198
Hazard Identification and Determination of Acceptable Levels	FS1070	197
Hygiene of Foodstuffs – European Parliament Regulation (EC) number 852/2004	FS1070	200

I

Document / Section Title	Section ID	Page
Identification and Segregation of Nonconforming Product	FS1150	282
Identification of Food Safety Records	FS1010	143
Implementing a HACCP Plan	FS1080	210
Implementing Prerequisite Programs	FS1050	173
Initiating a Corrective Action	FS1170	334
Initiation, Product Recall	FS1190	352
Inspection	FS1110	250
Inspection, Final	FS1120	259
Intended Use	FS1060	189
Internal Audit Follow-Up	FS1160	296
Internal Audit Planning	FS1160	293
Internal Audit Process, Validation of	FS1160	296
Internal Audit Program	FS1160	292
Internal Audit Reporting	FS1160	295
Internal Audit, Conducting	FS1160	293
Investi ting the Cause (Corrective Action)	FS1170	335
ISO/FDIS 22000:2005 – Food Safety Management Systems / Requirements for Any Organization in the Food Chain	FS1160	296
ISO/EN 19011:2002 – Guidelines for Quality and/or Environmental Management Systems Auditing	FS1160	296

J

Document / Section Title	Section ID	Page
Job Description Approval and Distribution	FS1040	166
Job Description form (FS1040-1)	FS1040	169
Job Description Format and Content	FS1040	164
Job Description Preparation	FS1040	164
Job Description Review	FS1040	166

K

Document / Section Title	Section ID	Page
Kitting Work Orders	FS1120	258

Document / Section Title	Section ID	Page

L

Labeling / Traceability System Development	FS1130	264
Labeling / Traceability System Review	FS1130	266
Labeling / Tracing Food Items and Ingredients	FS1130	264
Labeling Act, Consumer Packaging and (Canada – R.S. 1985, c. C-38)	FS1130	267
Labeling and Traceability – General	FS1130	264
Labeling, Food (21 CFR 101, USA)	FS1130	267
Labeling, Law Concerning Prevention of Unfair Gift and Unfair (Japan)	FS1130	267
Labeling, Packaging and	FS1120	259
Log, Vendor Performance (FS1100-4)	FS1100	247
Lot Identification / Product Traceability Log (FS1130-1)	FS1130	269

M

Maintenance, Storage, Handling (monitoring / measuring equipment)	FS1140	273
Management Planning	FS1020	152
Management Responsibilities and Authorities	FS1020	152
Management Review	FS1020	153
Manufacturing, Packing, or Holding Human Food – Current Good Manufacturing Practice in (21 CFR 110, USA)	FS1120	261
Matching, Recordkeeping and	FS1090	221
Measuring (Monitoring and) - General Requirements	FS1140	272
Monitoring and Measuring - General Requirements	FS1140	272

N

New Employee Orientation	FS1030	156
New Employee Selection	FS1030	155
New Vendor Evaluation	FS1100	233
Nonconformance Report	FS1150	283
Nonconformance Report form (FS1150-1)	FS1150	287
Nonconforming Product, Identification and Segregation of	FS1150	282
Nonconformity Report (FS1170-1)	FS1170	337
Nonconformity Reports	FS1170	333
Notification form, Approved Vendor (FS1100-3)	FS1100	245
Nutrition Improvement Law (Japan)	FS1130	267

O

Ongoing Training	FS1030	157
Operational Prerequisite Programs, Establishing	FS1050	174
Order Determination and Requisition	FS1090	219
Order Placement	FS1090	220
Out-of-Tolerance Conditions	FS1140	274

P

Packaging and Labeling	FS1120	259
Packaging and Labeling Act, Consumer (Canada – R.S. 1985, c. C-38)	FS1130	267
Planning, Management	FS1020	152

Document / Section Title	Section ID	Page
Planning, Prerequisite Program	FS1050	173
Preparation, Job Description	FS1040	164
Prerequisite Program Example (FS1050-1)	FS1050	177
Prerequisite Program Planning	FS1050	173
Prerequisite Programs, Implementing	FS1050	173
Prerequisite Programs, Operational, Establishing	FS1050	174
Prerequisite Programs, Reviewing	FS1050	174
Preventing Recurrence (Corrective Action)	FS1170	335
Prevention of Unfair Gift and Unfair Labeling, Law Concerning (Japan)	FS1130	267
Procedure and Work Instruction Format	FS1000	132
Process Steps, Control Measures, Flow Diagrams	FS1060	189
Product Characteristics	FS1060	188
Product Recall Communications	FS1190	352
Product Recall Initiation	FS1190	352
Product Recall Process, Review of	FS1190	353
Production	FS1120	258
Products of Animal Origin Intended for Human Consumption, Organization of Official Controls on (EC #854/2004)	FS1110	251
PRP Log Example (FS1050-5)	FS1050	185
PRP Planning	FS1050	173
PRPs (Operational), Establishing	FS1050	174
PRPs, Implementing	FS1050	173
PRPs, Reviewing	FS1050	174
Purchase Order Follow-Up (FS1090-4)	FS1090	229
Purchase Order form (FS1090-2)	FS1090	225
Purchase Order Log (FS1090-3)	FS1090	227
Purchase Requisition form (FS1090-1)	FS1090	223
Purchasing Review	FS1090	221

R

Document / Section Title	Section ID	Page
Recall (Food) Checklist (FS1190-1)	FS1190	355
Recall (Product) Communications	FS1190	352
Recall (Product) Initiation	FS1190	352
Recall Process (Product), Review of	FS1190	353
Recalled Product, Handling	FS1190	353
Receiving	FS1110	249
Receiving and Inspection Report (FS1110-2)	FS1110	255
Receiving and Inspection Review	FS1110	251
Receiving Log (FS1110-1)	FS1110	253
Record (Food Safety) Generation	FS1010	144
Record (Food Safety) Maintenance	FS1010	144
Recordkeeping and Matching	FS1090	221
Reevaluation, Vendor	FS1100	234
Regulation 306, Bioterrorism Act of 2002 (USA)	FS1130	267
Regulation (EC) #178/2002 of the European Parliament – General Food Law	FS1130	267
Regulation (EC) #852/2004 of the European Parliament of the Hygiene of Foodstuffs	FS1070	200

Document / Section Title	Section ID	Page
Regulation (EC) #854/2004 of the European Parliament – Organization of Official Controls on Products of Animal Origin Intended for Human Consumption	FS1110	251
Rejection, Discrepancies, and Disposition	FS1110	250
Release, Final (product)	FS1120	260
Request for Document Change form (FS1000-1)	FS1000	137
Requirements, Vendor Classification and	FS1100	232
Requisition, Order Determination and	FS1090	219
Responding to an Emergency	FS1200	362
Responsibilities and Authorities, Management	FS1020	152
Returned Goods	FS1150	284
Returned Goods Authorization form (FS1150-2)	FS1150	289
Review (Receiving & Inspection)	FS1110	251
Review of Product Recall Process	FS1190	353
Review, HACCP Plan	FS1080	210
Review, Purchasing	FS1090	221
Reviewing Prerequisite Programs	FS1050	174
Revision, HACCP Plan	FS1080	211
Risk Assessment and Evaluation	FS1200	359
Risk Management Solutions Test Report (FS1200-1)	FS1200	367
S		
Second-Party Vendor Audit Program	FS1100	235
Segregation (Identification and) of Nonconforming Product	FS1150	282
Selection and Assessment of Control Measures	FS1070	198
Software, Test (monitoring / measuring)	FS1140	275
SOP Form - Example (FS1050-2)	FS1050	179
Standard Operating Procedure (SOP) Form - Example (FS1050-2)	FS1050	179
Stocking	FS1110	250
Storage Map Example (FS1050-4)	FS1050	183
Storage, Handling, and Maintenance (monitoring / measuring equipment)	FS1140	273
Subcontractor Calibration, Control of	FS1140	275
Survey, Vendor - form (FS1100-2)	FS1100	239
T		
Taking Corrective Action	FS1170	335
Temporary Changes to Documents	FS1000	132
Test Software (monitoring / measuring)	FS1140	275
Traceability (Labeling and), System Development	FS1130	264
Traceability (Labeling and), System Review	FS1130	266
Traceability, Labeling and – General	FS1130	264
Tracing (Labeling) Food Items and Ingredients	FS1130	264
Training Log, Food Safety (FS1030-2)	FS1030	161
Training Requirements List, Food Safety (FS1030-1)	FS1030	159
Training, Ongoing	FS1030	157
U		
Unfair Gift and Unfair Labeling, Law Concerning Prevention of (Japan)	FS1130	267

Document / Section Title	Section ID	Page
Unsafe Products, Disposition of Potentially	FS1150	284
U.S. Americans with Disabilities Act of 1990	FS1040	167
U.S. Bioterrorism Act of 2002, Regulation 306	FS1130	267
U.S. Code of Federal Regulations, Title 9, Chapter 3, Part 417 – Hazard Analysis and Critical Control Point (HACCP) Systems	FS1080	212
U.S. Public Health Security and Bioterrorism Preparedness and Response Act of 2002	FS1010	145
Use, Intended	FS1060	189
V		
Validation of the Internal Audit Process	FS1160	296
Variables Control Chart (FS1180-1)	FS1180	349
Vendor (Approved) List - form (FS1100-1)	FS1100	237
Vendor (New) Evaluation	FS1100	233
Vendor Classification and Requirements	FS1100	232
Vendor Disqualification	FS1100	234
Vendor List, Approved (procedure)	FS1100	232
Vendor Performance Log (FS1100-4)	FS1100	247
Vendor Reevaluation	FS1100	234
Vendor Survey Form (FS1100-2)	FS1100	239
Verification (Hazard Analysis Preparation)	FS1060	190
Verification and Closure (Corrective Action)	FS1170	335
W		
Work Orders, Kitting	FS1120	258

[This page intentionally left blank.]

LaVergne, TN USA
25 March 2010
177176LV00004B/1/P

9 781931 591430